Shaima Almasry
Amr Elfayomy
Magda Eldomiaty

Placental analysis of idiopathic growth restricted pregnancies in KSA

AF153082

Shaima Almasry
Amr Elfayomy
Magda Eldomiaty

Placental analysis of idiopathic growth restricted pregnancies in KSA

LAP LAMBERT Academic Publishing

Impressum / Imprint

Bibliografische Information der Deutschen Nationalbibliothek: Die Deutsche Nationalbibliothek verzeichnet diese Publikation in der Deutschen Nationalbibliografie; detaillierte bibliografische Daten sind im Internet über http://dnb.d-nb.de abrufbar.
Alle in diesem Buch genannten Marken und Produktnamen unterliegen warenzeichen-, marken- oder patentrechtlichem Schutz bzw. sind Warenzeichen oder eingetragene Warenzeichen der jeweiligen Inhaber. Die Wiedergabe von Marken, Produktnamen, Gebrauchsnamen, Handelsnamen, Warenbezeichnungen u.s.w. in diesem Werk berechtigt auch ohne besondere Kennzeichnung nicht zu der Annahme, dass solche Namen im Sinne der Warenzeichen- und Markenschutzgesetzgebung als frei zu betrachten wären und daher von jedermann benutzt werden dürften.

Bibliographic information published by the Deutsche Nationalbibliothek: The Deutsche Nationalbibliothek lists this publication in the Deutsche Nationalbibliografie; detailed bibliographic data are available in the Internet at http://dnb.d-nb.de.
Any brand names and product names mentioned in this book are subject to trademark, brand or patent protection and are trademarks or registered trademarks of their respective holders. The use of brand names, product names, common names, trade names, product descriptions etc. even without a particular marking in this work is in no way to be construed to mean that such names may be regarded as unrestricted in respect of trademark and brand protection legislation and could thus be used by anyone.

Coverbild / Cover image: www.ingimage.com

Verlag / Publisher:
LAP LAMBERT Academic Publishing
ist ein Imprint der / is a trademark of
OmniScriptum GmbH & Co. KG
Bahnhofstraße 28, 66111 Saarbrücken, Deutschland / Germany
Email: info@lap-publishing.com

Herstellung: siehe letzte Seite /
Printed at: see last page
ISBN: 978-3-659-82406-7

Expression of tumor necrosis factor-α and localization of apoptotic bodies in placenta of normal and idiopathic intrauterine growth restriction cases in Al-Madinah Al-Munawwarah Province

By

Dr. Shaima Mohamad Ahmad Almasry
Dr. Amr Kamel Elsaid Elfayomy
Prof. Magda Ahmad Mohamed El-Domiaty
Prof. Fawzia Ahmed Habib Allah Babay
Prof. Maha Diaa-Eldin Safwat

This article contains the results of studies and research Supported by King Abdulaziz City for Science and Technology grant No. AT-29/230

1

Table of contents

SUMMARY

The delivery of a low birth weight infant as a result of intrauterine growth restriction (IUGR) represents a leading risk factor for neonatal morbidity and mortality.

Inflammatory cytokines such as tumor necrosis factor-alpha (TNF-α) and interleukin 6 (IL-6) are notorious for producing endothelial dysfunction. It has been reported that hypoxia could lead to elaboration of these powerful cytokines that might contribute to programmed or activated cell death in this organ.

This work analyzed cases with idiopathic IUGR from many points of view trying to reach proposed mechanisms for development of these cases aiming to propose a new management and/or further points for research concerning these cases.

Research Objectives

- Evaluate the maternal circulating levels of AM, NE, TNF-α and IL-6 in idiopathic IUGR cases in comparison to appropriate for gestational age (AGA) cases.

- Identify the correlation of the inflammatory cytokines; TNF-α and IL-6 to AM.

- Assess the gross morphology of placentae and the histomorphometric characteristics of the placental terminal villi and their respective villous capillaries.

- Studying the histopathological changes in the placental stem villi and their vessels.

- Identifying the correlation of the above placental changes to the neonatal birth weight and selected morphological features of the placenta aiming to reach the pathogenesis underlying the development of idiopathic IUGR.

Research design and methodology

- About 50 placentas were obtained from normal vaginal and cesarean section delivery of idiopathic IUGR. The control group was consisted of 25 normal placentae from mothers having fetuses appropriate for gestational age.

- 5 ml of blood samples were collected from every participant under complete aseptic condition for assessment of maternal serum levels of AM, NE, TNF-α and IL-6.

- Placental tissue specimens were processed for hematoxylin-eosin staining for morphometric and histopathological evaluation of the chorionic villi using image analysis system.
- Immuno-histochemical techniques were used for evaluation of TNF-α expression and detection of apoptotic bodies in the trophoblastic tissue.
- Statistical analysis of all obtained results of examination and analysis.

Results

- Maternal serum analysis reveals that women with idiopathic IUGR show a significantly higher serum levels of AM, TNF-α and IL-6 than those of the AGA group but the serum level of NE was insignificantly higher among IUGR group. In idiopathic IUGR cases, there is a significant correlation between plasma levels of AM and TNF-α but not between plasma levels of AM and IL-6.
- Histopathological study of placental stem villi reveals villitis and the degenerative changes in the villi that are significantly higher in idiopathic IUGR cases than in control ones. Positive correlation is detected between the fetal birth weight and different pathologic features in the stem villi as stem artery number, arterial narrowing, villitis and stem villous degenerative changes.
- Morphometric analysis of placental terminal villi reveals that idiopathic IUGR is associated with reduced growth of terminal terminal villi accompanied by changes in measures of villous capillarization as compared with control placentae. Significant positive correlation is detected between birth weight and terminal villous capillary number.
- Analyzing placental tissue TNF-α determines its localization in both deciduas and chorionic villi with mean area percent of TNF-α immunostainaing significantly higher in the idiopathic IUGR group compared to the control one. Importantly, TGCs of control specimens show deficient or negative TNF-α immunoexpression while those of IUGR group show positive staining.
- Studying apoptosis in placental tissue shows that apoptosis is more abundant in the trophoblasts and proves that the rate of apoptosis is significantly higher among placental tissues of pregnancies complicated with idiopathic IUGR than in placentae of normal uncomplicated pregnancies.

Conclusions

- These results could raise the hypothesis that the stem villi more than the terminal villi could represent the mystery for the development of idiopathic IUGR. The decreased number of stem arteries and or their narrowing might be causative mechanisms for decreased birth weight and development of idiopathic IUGR in Saudi, also degeneration of stem villi and villitis of unknown etiology could be underlying mechanisms.

- Through this study, it is proposed that the enhanced rate of apoptosis in the trophoblast may have an important role in the pathogenesis of IUGR. The increased maternal serum TNF-α and enhanced placental expression of this cytokine that has strong pro-inflammatory activity could represent the underlying mechanism for the development of villitis and the increased rate of apoptosis.

Recommendations

- Matrnal serum cytokine level can serve as useful biochemical markers for idiopathic IUGR and experimental studies should be done to ascertain the effect of increased TNF-α as inflammatory cytokine on the fetal development and the possible opposing effects of anti-inflammatory cytokines.

- The stem villi more than the terminal villi should create a center of attention point for research to ascertain the causative mechanism for IUGR. Also the trophoblastic giant cells should be experimented upon to ascertain its role in cytokine secretion.

INTRODUCTION

The placenta is the vital organ for maintaining pregnancy and promoting normal fetal development. The intrauterine existence of fetus is dependent on the placenta. The one and most important cause of neonatal loss is low birth weight. Intrauterine growth restricted (IUGR) or Small-for-Date (SFD) babies are a group of low birth weight babies having impaired growth rate. IUGR affects a fair number of newborn infants worldwide, mainly in developing countries. The total incidence of IUGR was 13.3% and 8.1% was categorized as unexplained IUGR (Villar et al., 2005). De Onis et al. (1998) estimated that at least 13.7 million babies in developing countries are already malnourished at birth (IUGR-Low birth weight) every year, representing 11% (ranging from 1.9% to 20.9%) of all newborns in these countries. The incidence of IUGR-LBW is about 6 times higher in developing than in developed countries (Villar et al., 1994). In Saudi Arabia and in other developing countries, the educational level, physical activity, annual family income and living accommodation of pregnant women were found to be statistically significant to cause IUGR.

The etiologies of IUGR of fetuses are numerous, but in some cases, there are no obvious fetal or maternal causes. The placentas of these "idiopathic" intrauterine growth retarded babies might give a clue to the etiology of the growth retardation. The contribution of placental changes to the pregnancies resulting in birth of intrauterine growth restricted fetuses remains controversial (Biswas and Ghosh, 2008).

The process of implantation and placentation requires the production of a plethora of factors. The tumor necrosis factor (TNF) gene family is believed to participate in these processes via regulation of genes involved in apoptosis (programmed cell death) as well as other critical placental functions such as hormone production (Hunt et al., 1996). TNF is among the apoptosis-inducing ligands that are transcribed and translated in human placentas (Chen et al., 1991; Phillips et al., 2001; Gill et al., 2002; Gill and Hunt, 2004).

Apoptosis is a physiological process in development, tissue homeostasis, and disease. Apoptosis is a normal event in several reproductive tissues including human placenta (Choi et al., 2003). Apoptosis occurs in the villous trophoblast of normal placentas throughout pregnancy, but with higher frequency near term. In pregnancies complicated by IUGR, a greater incidence of villous and extravillous trophoblast apoptosis has been observed, suggesting that the deregulation of trophoblast apoptosis may contribute to pathological conditions.

DESIGN & METHODOLOGY

Subjects details

The cases were chosen from 2 University-teaching hospitals; "Al-Madinah Maternity and Children Hospital" and Ohoud Hospital", between April, 2010 and March, 2011. This case-control study was approved by "Local Medical and Health Sciences Research Committee". Full-term freshly delivered placentae were collected both after normal deliveries and Caesarean sections. 50 placentae were associated with idiopathic IUGR singleton babies and 25 were gestation-matched control pregnancies from singleton normal weight babies. Control cases were selected to match IUGR cases according to gestation. Placentae were obtained from consenting pregnant women. To eliminate the confounding effects of premature birth, only term (≥37weeks) cases were chosen for this study. Gestation for both IUGR and control women was calculated from the last menstrual period dates and confirmed by an ultrasound examination performed 11-13 weeks gestation (Hadlock et al. 1992).

The inclusion criteria for IUGR cases were; serial ultrasound fetal weight below the 10th percentile for gestational age (Hadlock et al. 1992) and any two of the following criteria diagnosed on antenatal ultrasound; abnormal umbilical artery Doppler flow velocimetry, oligohydramnios as determined by amniotic fluid index (AFI) <5 (Cunningham et al. 2010) or asymmetric growth of the fetus as quantified from the head circumference to abdominal circumference ratio. Growth restriction was confirmed at birth if neonatal weight was less than the 10th percentile (Williams et al. 1982; WHO, 1995).

The exclusion criteria for both IUGR and control cases were multiple pregnancies, maternal smoking, preeclampsia, prolonged rupture of the membranes, placental abruption, intrauterine viral infection, fetal congenital anomalies, genetic syndromes, maternal autoimmune diseases and diabetes. Thus our IUGR population is considered idiopathic.

Ethical issues

Prior to conduct the study, approval of the research protocol was obtained from the Medical Officers in Charge of the Antenatal clinics before the study commenced.

Participation in the study was optional and written informed consent was obtained from participating women. Local Ethics Committees approved the study. A comprehensive questionnaire was administered to all individuals who consented to participate in the study to show complete history of the case and suspected environmental or genetic factors that might cause IUGR.

Blood Samples

5 ml of blood samples were collected from every participant during the first stage of labor, or before receiving anesthesia in cases of elective caesarean section under complete aseptic condition and left for 30-60 minutes for spontaneous clotting at room temperature then centrifuged at 3000 rpm for 10 minutes. Serum samples were separated into another set of tubes and kept frozen at - 80°C for determination of maternal serum level of AM, TNF-α, IL-6 and NE.

Detection of TNF-α, AM, IL-6 and NE in the maternal plasma using Enzyme Immunoassay (EIA) (Hailman et al., 1996)

1. Prepare working solutions of the (TNF-α, AM, IL-6 and NE)-HRP conjugate and wash buffer.

2. Remove the required number of microwell strips. Reseal the bag and return any unused strips to the refrigerator.

3. Pipette 50μl of each calibrator, control and specimen sample into correspondingly labeled wells in duplicate.

4. Pipette 100 μl of the conjugate working solution into each well (We recommend using a multichannel pipette).

5. Incubate on a plate shaker (approximately 200 rpm) for 1 hour at room temperature.

6. Wash the wells 3 times with 300 μl of diluted wash buffer per well and tap the plate firmly against absorbent paper to ensure that it is dry (The use of a washer is recommended).

7. Pipette 150μl of TMB substrate into each well at timed intervals.

8. Incubate on a plate shaker for 10-15 minutes at room temperature (or until calibrator A attains a dark blue color for desired OD).

9. Pipette 50µl of stop solution into each well at the same timed intervals as in step 7.

10.Read the plate on a microwell plate reader at 450 nm within 20 minutes after addition of the stop solution.

Morphological examination of placenta

The placentae, the membranes were trimmed up to the margins of the placentae. The placentae were dried with blotting paper and the followings were recorded:

- Positions of insertion of the umbilical cords.
- Placental diameters along two axes, perpendicular to each other, by means of measuring tape. Then the average of the 2 diameters was calculated for statistical purposes.
- Placental co-efficient (= Placental weight in grams ÷ Birth weight in grams).
- Hematomas and fibrotic spots.

Tissue samples

Placental samples were obtained by standard procedure; one full depth cube of placental tissue from macroscopically normal placental disk, 5cm from the umbilical cord insertion (Fig. 1). Each cube of tissue was processed to wax using routine laboratory techniques. Paraffin blocks were cut at 4 mm thickness and were processed for hematoxylin-eosin (HE) staining, carried out according to conventional procedures.

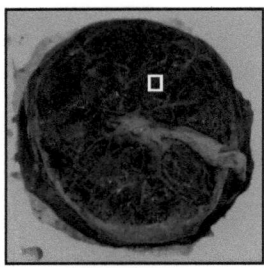

Fig. 1: Photomicrograph of normal placenta showing: excision of the cube like specimen from macroscopically normal placental disk, 5 cm from the umbilical cord insertion.

Haematoxylin and Eosin staining

The placental specimens were fixed for a minimum of 24 h in 10% buffered formalin and then will be processed through alcohol (90%, 100%), chloroform and wax, in a Reichert automatic tissue processor before embedding in wax. The histological sections were cut from each block, each 5 μm in thickness. Sections were dewaxed with xylene and rehydrated through a decreasing alcohol series. Five-micrometer adjacent sections were stained with Harris's HE. All sections were examined for the structure of the placental tissues and photographed.

Placental analysis

Morphometric measures were taken from 75 blocks: 50 from idiopathic IUGR and 25 from control placentae. The same investigator performed all analyses blind. Morphological measurements were taken using an Image Analyzer (Leica Q Win standard, digital camera CH-9435 DFC 290, Jermany). Five fields/slide counted, avoiding areas of placental infarction; areas of intervillous fibrin deposition and histological artifact (Giles et al., 1985). The technical possibilities offered by image-analysis system allowing visual control and simultaneous statistical processing of the measurement data, these make it possible to register complex quantitative data concerning the characteristics features of the terminal villi.

Terminal villous refers to villi <80 μm in diameter and embraces both terminal and smaller mesenchymal villi and mature intermediate villi (Mayhew, 2002). The terminal villi and villous capillaries found totally inside the microscopic field (area=786432.0 μ2) were used for group comparison. The total and mean areas (μ2) were measured and the number of terminal villi and villous capillaries were counted (Giles et al. 1985). The mean areas of villi and villous capillaries were calculated by the ratio between the total area and number of villi or vessels in the field. Capillarization index (%) was defined as the total area of villous capillaries and terminal villi ratio (vascular total area/villous total area x100) (Mayhew et al. 1994) (Fig. 2).

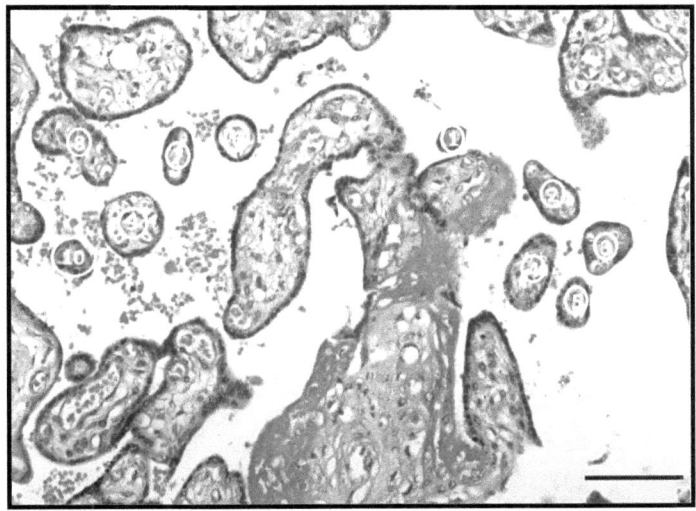

Fig. 2: Photomicrograph of placental terminal villi and villous vessels. Terminal villi (<80 μm in diameter) totally inside the field and uniform for morphometric analysis; terminal villous total area = 1+2+3+4+5+6+7+8+9+10; villi mean area = (1+2+3+4+5+6+7+8+9+10)/10 (HE, x200; bar: 100 μm).

Histopathological examination

The intervillous spaces of placentae of control and idiopathic IUGR cases were examined. Complete cross sectioned encircled villi with core of connective tissue were selected for examination of the following: I. the stem villi for; a) the number of stem arteries. B) any pathological findings in the stem arteries as degeneration, herniation, dissection of the wall and or thrombosis. C) pathological changes in the interstitium of the stem villi as hyalinization or fibrosis. II. the terminal villi for fibrotic changes.

For the statistical study, examination was performed in randomly selected 5 different high power (x200) fields of each specimen. The score in Table 1 was used to determine some pathological features in both IUGR and control cases.

Table 1: The score for determination of some pathological features in placentae of IUGR and control cases

The pathological feature	The score	
Narrowing of the stem arteries	In more than 2 fields	Positive
	In less than 2 fields	Negative
Villitis of stem villi (presence of cellular infiltration)	In 2 fields or more ± degenerative changes in the stem villous	Positive
(low-grade lesions affecting less than 10 villi per focus were excluded)	In less than 2 fields	Negative
Terminal villous fibrosis	In more than 2 fields	Positive
	In less than 2 fields	Negative

Immunohistochemical staining of TNF-α

3 μm-thick sections were cut by microtome and collected onto poly-L-lysine-coated slides and analyzed using an immunoperoxidase technique. Briefly, after the slides were deparaffinized, they were immersed in phosphate-buffered saline for 10 minutes at room temperature, and then treated with 0.3% H_2O_2 in absolute methanol for 20 minutes at room temperature to block endogenous peroxidase activity. After PBS rinse, immunostaining were performed using mouse monoclonal anti-human TNF-α antibodies (purified MAb-IgG1-9F274) (0.5 mg/ml), (USBiological, T9160-

21B). To localize binding of the primary antibody, the slides were incubated with the secondary anti-mouse antibodies universal kits obtained from Zymed Corporation. Sections incubated without the primary antibody were included as negative controls in all experiments (Kiernan 1999). Then, sections incubated in DAB reagent and counterstained with hematoxylin and cover slipped using Protex mounting media (DAB-Stock Stain box; Boster Biotechnology).

Evaluation of TNF-α immunoexpression

Tissue sections were examined using an optical microscope at 200 x magnification for the initial screening. Measurements were performed at 400 x magnification. For each specimen, 5 high power fields were randomly selected, photographed and stored. The Digitalized pictures were examined by 2 investigators on a high-resolution color display. A positive reaction for TNF-α protein was detected as a brown yellow granulation in the cellular membranes and cytoplasm. Negative cells had a clear cell structure without brown granulation in their cellular membrane and cytoplasm (Yu et al., 2007). The area percent of the TNF-α immunoreaction was measured using the Image Analyzer (Leica Q Win standard, digital camera CH-9435 DFC 290, Jermany) through making use of the technical possibilities offered by image-analysis systems that allows visual control and simultaneous statistical processing of the measurement data and make it possible to register complex quantitative data concerning the area percent of standardized degree of brown immunostaining for all the slides.

Terminal Deoxynucleotidyl Transferase (TdT)-mediated dUTP nick end labeling immunohistochemistry for apoptosis; Staining Procedure (Recommended by Ventana Company):

1. Load slides, antibody, and iView™ detection kit dispensers onto BenchMark instrument.
2. Select CC1 Standard pretreatment.
3. Antibody incubation should be set for 32 minutes at 37° C.
4. Start the run.
5. When the staining run is complete, move slides from instrument and rinse well with wash buffer.

6. Cover slip.

The immunohistochemical approach to apoptosis involves detection of digoxigenin-labeled genomic DNA in fixed tissues by an immunoperoxidase. The labeled target is the multitude of 3'OH DNA ends produced by DNA fragments, a hallmark of apoptosis. These fragments typically are localized in morphologically identifiable nuclei and apoptotic bodies. Apoptotic bodies were marked by brown to black precipitates and graded as: 0, no staining; 1+, less than 50% staining in the cells; 2+, 50% or more staining in the cells (Rivera et al. 1998; Sur et al. 2007).

Computer-assisted image analyzing study

Computer-assisted analysis was performed as previously described (Bocci et al., 2001). Images were digitized in a 512×512-pixel matrix, using a color video camera and a microcomputer processor. Digitalized pictures were visualized on a high-resolution color display. The true color image analysis software package using image analysis system (Leica Imaging System, Switzerland & Germany) was run for manipulation, quantification of the images and data collection.

Computerized imaging analysis systems have been introduced in order to minimize subjectivity in quantifying the positive TNF-α immunohistochemical stained cells and the apoptotic bodies counting. Also, it was used for counting and measuring areas of the chorionic villi and chorionic villous vessels in five randomly selected different fields.

Statistical analysis

Statistical analysis was performed using SPSS (version 13) statistical package. The demographic and immunohistochemical data were expressed as mean ± SEM. Significance of differences between groups were calculated using independent student t-test and Mann-Whitney U-test.

RESULTS AND DISCUSSION

Clinical criteria of the cases

Clinical characteristics of the studied groups are shown in Table (2). Women in the control and idiopathic IUGR groups has mean age of 29.75y and 28.64y respectively with no significant effect of age on IUGR (p=0.264). There are no significant differences between both groups regarding the parity (p=0.062), the mode of delivery (p=0.297) or the sex of newborns (p=0.865). Significant difference between both groups was found regarding the fetal birth weight (p=0.000).

Table 2: Demographics and general obstetric outcomes in the studied groups

Variables	Idiopathic IUGR group (n=50)	Control group (n=25)	P value
Maternal age (years) (mean ± SEM)	28.64±0.58	29.75±0.79	0.264
Gestational age (weeks)	38.11±0.13	38.88 ±0.19	0.001[*]
Parity (no. and %)			
Primiparae	34 (68%)	16 (32%)	0.062
Multiparae	22 (88%)	3 (12%)	
Mode of delivery (no. and %)			
Vaginal delivery	32 (64%)	19 (76%)	0.297
Cesarean section	18 (36%)	6 (24%)	
Sex of newborn (no. and %)			
Male	17 (34%)	9 (36%)	0.865
Female	33 (36%)	16 (64%)	
Birth weight (gm)	2203.59±39.	3396.47±62	0.000[*]

(mean ± SEM)	08	.43	

Data are presented as mean ± standard error of means (SEM). Significance was taken as P<0.05 for Independent Samples-t test (*). IUGR: intrauterine growth restriction.

A. ANALYSIS OF THE SERUM LEVEL OF AM, NE AND INFLAMMATORY CYTOKINES; TNF-α AND IL-6

As shown in Table (3) and Diagram (1), women with idiopathic IUGR have a significantly higher serum levels of AM (p=0.008), TNF-α (p=0.016) and IL-6 (p=0.029) in idiopathic IUGR group than those of the control group. The serum level of NE was higher among IUGR group as compared with the control group but the difference in did not reach a significance level (p=0.269).

Interestingly, there is a significant correlation between plasma levels of AM and TNF-α (r=0.417, p=0.003) in idiopathic IUGR cases. Whereas, there is no significant correlation between plasma levels of AM and IL-6 (Diagrams 2, 3).

Table 3: Mean levels of AM, NE, TNF-α and IL-6 in maternal plasma of normal deliveries and IUGRs

Variable	Idiopathic IUGR group (n=50)	Control group (n=25)	P value
AM (pg/ml) (mean ± SEM)	64.21±3.12	49.54± 4.27	0.008*
NE (pg/ml) (mean ± SEM)	8.26 ± 1.07	6.39 ± 1.29	0.294
TNF-α (pg/ml) (mean ± SEM)	4.27±0.37	2.94± 0.13	0.016*
IL-6 (pg/ml) (mean ± SEM)	64.93±7.16	40.26±6.29	0.029*

Data are expressed as mean ± SEM. *Significance was taken as P<0.05 for Independent Samples-t test. AM: adrenomedullin; NE: Norepinephrine; TNF-α: tumor necrosis factor-alpha; IL-6: interleukin-6; IUGR: intrauterine growth restriction.

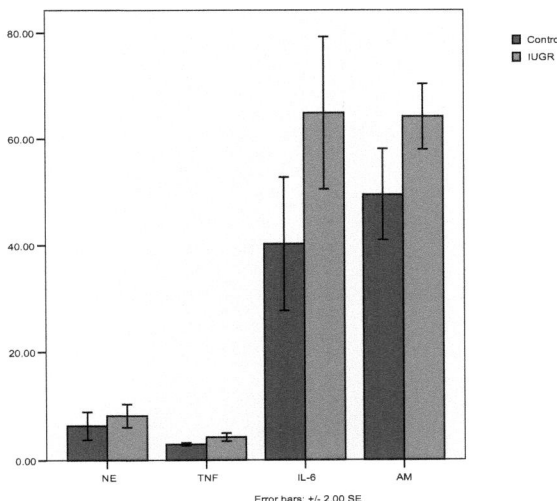

Diagram 1: Data represents mean +/- SEM maternal serum levels of NE (norepinephrine), TNF-α (tumour necrosis factor-alpha), IL-6 (interleukin-6) and AM (adrenomedullin). Significance of difference between groups at p<0.05.IUGR: intrauterine growth restriction.

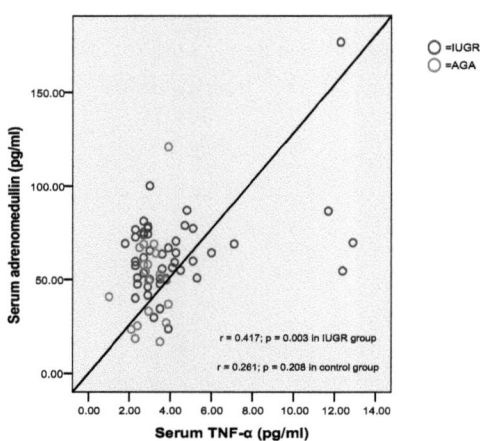

Diagram 2: Correlation analysis of AM to TNF-α in maternal plasma of normal deliveries and IUGRs.

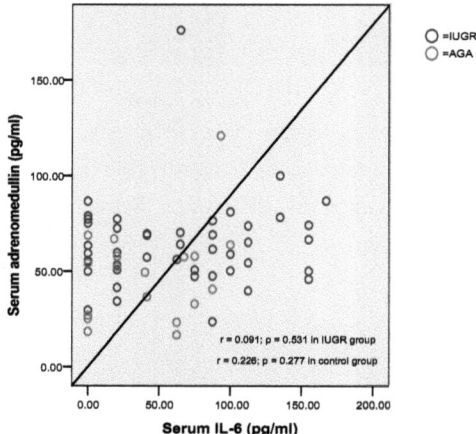

Diagram 3: Correlation analysis of AM to IL-6 in maternal plasma of normal deliveries and IUGRs.

DISCUSSION

TNF-α biochemical finding

A main finding in this work from Saudi Arabia is that the TNF-α plasma level is up-regulated significantly in the plasma of the idiopathic IUGR cases. Up-regulation of TNF has been postulated to be a survival mechanism in the IUGR fetus, by inducing muscle insulin resistance, thus enabling glucose to be spared for brain metabolism (Briana and Malamitsi-Puchner, 2009). This is in agree with Heyborne et al., (1992) and Stallmach et al., (1995) who found abnormally high TNF-α levels in amniotic fluid in patients with IUGR. Besides, Holcberga et al., (2001) used a perfusion model system and found that, IUGR placentae may have the capacity to release more TNF than normal placentae upon angiotensin II stimulation.

On the other hand, Schiff et al., (1994) and Opsjon et al., (1995) evaluated maternal and fetal plasma TNF-α level in pregnancies associated with small-for-gestational-age newborns. Both of them studied only newborns with idiopathic growth retardation and decreased (Schiff et al., 1994) or normal (Opsjon et al., 1995) plasma TNF-α level were found. Seremak-Mrozikiewicz et al., (2008) added that increased umbilical artery vascular impedance and signs of brain sparing could be related to IUGR and increased TNF-alpha level in maternal serumTNF-α is present in human fetal blood throughout pregnancy as well as in amniotic fluid and maternal blood (Vince et al., 1992). Several factors could explain the presence of this inflammatory cytokine in plasma of normal pregnant females e.g., the immunologic stimuli occurring as a part of adaptation of immune system to pregnancy, also, the local inflammatory responses initiated in the uterus at the implantation site due to the entry of the highly invasive trophoblast (Pijnenborg et al., 1998).

TNF concentrations seem to determine whether the cytokine exerts beneficial or harmful effects and there must be a complex interaction pattern between TNF concentration, tissue and cell type, TNF receptor distribution and duration of TNF stimulation leading to a specific physiological or pathological reaction (Haider and KnÖfler, 2009).

In the present work, histopathological analysis of IUGR placenta revealed villitis of unknown etiology (VUE) as a probable mechanism of IUGR and this support the finding of Raymond & Redline, (2007). This could be related to the increased maternal serum TNF-α level owing to its strong pro-inflammatory activity. Furthermore, elevated TNF-α could potentially promote endothelial cell activation which is also being discussed for IUGR (Johnson et al., 2002). Chen et al., (1996) and Meekins et al., (1994) added that, TNF plays a significant role in changing the balance between oxidant and antioxidant, the pattern of prostaglandin production and expression of adhesion molecules in blood vessels.

IL-6, AM and NE biochemical findings

In this study, the sympathoadrenal system in idiopathic IUGR was evaluated. Sympathoadrenal system function may be altered following IUGR. In this situation, there is decreased oxygen delivery to the fetus over a prolonged period being widely thought to be a principal causal factor, a view that has received recent experimental support (Boyle et al., 1996; Murotsuki et al., 1997). In this study, NE was increased among patients with IUGR, Alterations in their production by the sympathetic nervous system and adrenal medulla are considered to play a very important role in the adaptation of the fetus to reduced oxygen availability. Inagreement with this study, increased fetal plasma Epi and NE concentrations have been observed in experimental models of intrauterine growth retardation (Bassett and Hanson, 1998) as well as during fetal hypoxemia (Gagnon et al., 1997; Stonestreet et al., 1995), and it has generally been considered that the increases in catecholamine concentrations play an important role in the process of adaptation by the fetus to inadequate oxygen and metabolic substrate availability. These observations provide unequivocal evidence that the marked increases in Epi and NE concentrations observed in the fetus during hypoxemia or other adverse intrauterine nutritional conditions contribute directly to the disproportionate retardation of fetal development observed in these situations, irrespective of any limitation in substrate or oxygen supply (Bassett and Hanson, 1998).

Tinkanen et al. (1993) have found that NE level in plasma were significantly higher in patients with pre-eclampsia than in women with a normal pregnancy. Also, NE level in blood was significantly higher in patients with severe pregnancy induced hypertension than those in control subjects, but not in moderate cases. Levels of plasma NE was increased among patients with moderate PIH, but did not reach a statistical significance as compared with the control group (Zhang, 2001).

Adrenomedullin is a potent vasodilator peptide eliciting a long-lasting vasorelaxant action (Akturk et al., 2007). Because of its potential role in regulating systemic and placental blood flow, several authors have investigated AM production in IUGR, and controversial results have been reported. In this study, AM plasma Level was significantly higher in patients with IUGR than those in control subjects. However, maternal plasma level of AM was increased in IUGR than control groups with no statistical significant differences (Di Iorio et al, 2000; Di Iorio et al., 2003; Akturk et al., 2007). Our observations support the hypothesis that an increase in adrenomedullin secretion in IUGR may be a compensatory mechanism in fetoplacental ischemia or in impaired blood flow in the uteroplacental or fetal circulation. Also, Upton et al. (1997) have mentioned that AM has been known to reduce the contractile response of isolated rat uterus, which has AM binding sites. The physiologic increase of plasma AM during pregnancy is considered to suppress uterine contraction and increase uterine perfusion (Hinson et al. 2000).

In recurrent pregnancy loss, Nakatsuka et al. (2003) observed that AM plasma is increased. They supposed that this peptide may serve as a biochemical marker to identify women with recurrent pregnancy loss associated with impaired uterine perfusion. The plasma level of adrenomedullin is elevated in various diseases including hypertension, diabetes, septic shock, or systemic lupus erythematosus, which are often associated with pathologic processes of the vasculature. These reports suggest that plasma adrenomedullin may increase in compensation for vascular dysfunction (Jougasaki and Burnett, 2000; Cheung et al., 2000)

Cytokines are also thought to play an important role in placental development and growth, although they are poorly studied. The placenta produces pro-inflammatory cytokines, such as IL-6 and TNF-α (Dudley et al., 1992), and human decidua cells, in vitro, secrete IL-6, which increases markedly after stimulation with IL-1α , IL-1ß, and TNF-α (Meisser et al., 1999). In this study IL-6 was increased among patients with IUGR. Also, IL-6 has been found to be increased in women with IUGR (Street et al., 2006; Tosun et al., 2010), even though other studies have not confirmed these findings, reduced IL-6 levels have been documented in IUGR (Odegard et al., 2001). It could be speculated that, possibly due to hypoxia and/or nutrient deficiency in the former and to, impaired trophoblast function and severe placental insufficiency in the latter supporting the hypothesis that IL6 may be related to fetal growth in the feto–

maternal interface. However, Bartha et al. (2003) have found that the serum levels of IL-6 were similar in the studied groups. In humans, pre-eclampsia is also characterized by both growth factor and cytokine modifications. Several cytokines, such as TNF-α and IL-1ß, are increased in the placenta, (Grimble, 2002; Tosun et al., 2010).

It may be hypothesized that placental insufficiency in cases of IUGR could be caused by an immunological phenomenon. Elevation of IL-6 cytokine could be a specific phenomenon of certain subsets of IUGR identifying cases of placental dysfunction.

Cytokine regulation of AM has been extensively studied in various animal cells (Hinson et al., 2000), but no data are available regarding its regulation in the placenta. Because of the possible roles of these cytokines in placental biology and pathophysiology, we wished to determine their effects on AM secretion. Li et al. (2003) have studied this effect in preeclampsia. Hypoxia is considered to be a likely etiologic factor in PE (Roberts et al., 2001; Marinoni et al., 2011). It is also known that AM is likely a hypoxia-induced gene (Cormier-Regard et al., 1998). Also, in IUGR, there is hypoxia and increased apoptosis, characteristics of the disease (Huppertz et al., 1998; Huppertz et al., 1999). This might explain the positive correlation between AM and IL-6 cytokine in the present study.

The discrepancies between studies may be explained by the number of cases, the sample selection, or by using different immunoassay system in different studies. These factors must be taken into account while planning future studies. In-addition, IUGR is a heterogeneous condition that includes a wide variety of situations ranging from physiological small-for-gestational-age babies to abnormal conditions, including cases of fetal malformations, infections or placental insufficiency. This could explain the differences in the results of previous studies. However, further studies including a larger number of cases could help to definitively clarify the relationships between AM, EP, or IL-6 and placental insufficiency and to evaluate the predictive value of these biological markers in the diagnostic and therapeutic situations.

In conclusion, Idiopathic IUGR pregnancies in Saudi Arabia are associated with increased the circulatory levels of the pro-inflammatory cytokine TNF-α in the maternal plasma. This might indicate a common pathway in the pathogenesis of

IUGR. Use animal models to investigate potential anti-inflammatory agents in adverse reproductive outcomes could be beneficial in supporting this hypothesis.

B. GROSS MORPHOLOGY OF THE PLACENTAE

The mean diameters of placentae from two axes are 16.3cmx15.17cm (average: 15.73±0.17) in idiopathic IUGR placentae and 17.69cm x 16.72cm (average: 17.20±0.23) in control placentae. The placentae associated with idiopathic IUGR are significantly (p=0.000) smaller in diameters than those of the control group. Birth weight is correlated significantly with the average placental diameters (r=0.619; p=0.000) (Table 4; Diagram 4).

The mean placental weights are significantly (p=0.000) lower in idiopathic IUGR newborns. Birth weight is correlated significantly with placental weights (r=0.650; p=0.000). Placental weights are also evaluated in relation to fetal weight through the placental co-efficient which is the ratio of placental weight to fetal weight (placental weight: fetal weight) and there is a significant (p=0.026) difference between both groups being higher in idiopathic IUGR and the placental co-efficient values are significantly correlated with the birth weights (r=-0.456; p=0.000) (Diagram 4).

Positions of insertion of umbilical cords and gross placental changes (fibrosis, hematomas) are stated in Table 4. In idiopathic IUGR placentae, positions of insertion of umbilical cords are found to be eccentric in 34, central in 7, marginal in 3, and velamentous in 6 out of 50 placentae. In control placentae, it is eccentric in 15, central in 7, marginal in 2, and velamentous in 1 out of 25 placentae. In majority of idiopathic IUGR (68%) and control (60%) cases, eccentric insertion of cord is noted with no significant difference could be detected between both groups (p=0.149).

Placental morphometric parameters

A. Terminal villi

In idiopathic IUGR pregnancies, the mean total area of terminal villi per villous cross-section is significantly lower (p=0.048), and the mean terminal villi number is significantly lesser (p=0.000) than in control pregnancies while, the mean values of mean villous area are nearly similar (14777.70±487.17µ2; 14887.69±576.23µ2; p=0.891) in idiopathic IUGR and control groups respectively (Table 5).

B. Villous vessels and capillarization index

There are significant differences between idiopathic IUGR and control placentae in the mean total capillary area (p=0.000) and in the mean number of capillaries (p=0.001) per villous cross-section while, the mean values of mean capillary and villous areas are nearly similar in both groups (p=0.983) (Table 6).

Table 4: Gross morphological examination of the placentas and mode of insertion of the umbilical cords

Variable	Idiopathic IUGR group (n= 50)	Control group (n =25)	P value
Placental diameter (cm) **(Mean of average ± SEM)**	16.30 * 15.17 (15.73 ± 0.17)	17.69 * 16.72 (17.20 ± 0.23)	0.000 *
Placental weight (gm) **(mean ± SEM)**	354.80 ± 9.82	488.08 ± 14.09	0.000 *
Placental co-efficient **(mean ± SEM)**	0.16 ± 0.005	0.14 ± 0.005	0.026 *
Insertion of the cord (no. and %) **Central** **Eccentric** **Marginal** **Velamentous**	7 (14%) 34 (68%) 3 (6%) 6 (12%)	7 (28%) 15 (60%) 2 (8%) 1 (4%)	0.149
Gross placental changes (no. and %) **Fibrosis**	16 (32%) 15 (30%)	2 (8%) 2 (8%)	0.023 ** 0.033

Hematomas			**

Data are presented as mean ± standard error of means (SEM). Significance was taken as P<0.05 for Independent Samples-t test (*) and Mann-Whitney test (**). IUGR: intrauterine growth restriction.

Table 5: Morphometric criteria of placental terminal villi in idiopathic IUGR and control groups

Terminal villi	Idiopathic IUGR (n= 50)	Control (n =25)	P value
Total area (μm2) (mean ± SEM)	431155.03 ± 32005.65	538226.12 ± 39759.26	0.048*
Mean area (μm2) (mean ± SEM)	14777.70 ± 487.17	14887.69 ± 576.23	0.891
Number (mean ± SEM)	29.44 ±1.88	70.06 ± 5.03	0.000*

Data are presented as mean ± standard error of means (SEM). Significance was taken as P<0.05 for Independent Samples-t test (*). IUGR: intrauterine growth restriction.

Table 6: Morphometric criteria of placental terminal villous capillaries in idiopathic IUGR and control groups

Terminal villous capillaies	Idiopathic IUGR (n= 50)	Control (n =25)	P value
Total area (µm2) (mean ± SEM)	106909.53 ± 10136.41	187205.94 ± 15900.59	0.000 *
Mean area (µm2) (mean ± SEM)	2508.73 ± 148.58	2513.79 ± 133.95	0.983
Number (mean ± SEM)	47.09 ± 4.44	73.35 ± 5.13	0.001 *
Capillarization index (%) (mean ± SEM)	27.62 ± 2.04	34.42 ± 1.98	0.038 *

Data are presented as mean ± standard error of means (SEM). Significance was taken as $P<0.05$ for Independent Samples-t test (*). IUGR: intrauterine growth restriction.

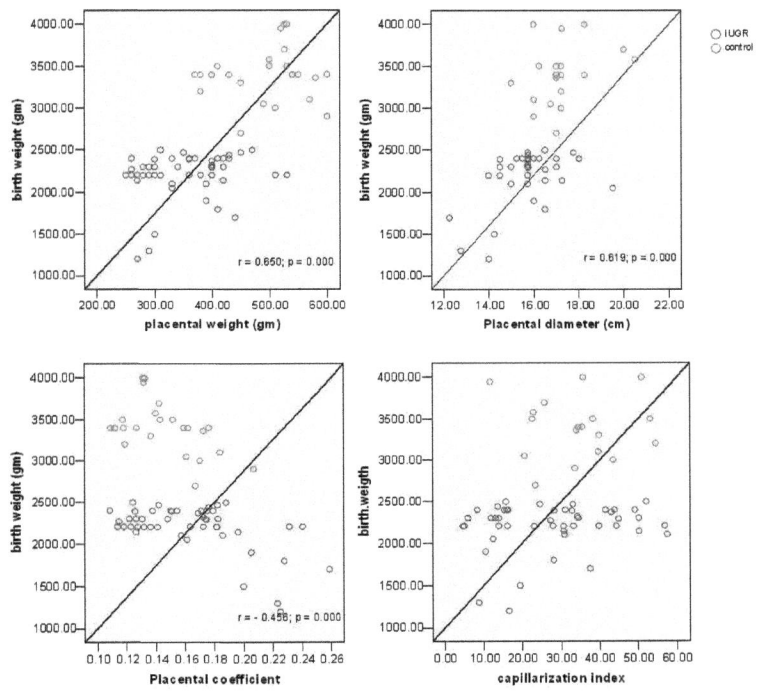

Diagram 4: Neonatal birth weight correlate significantly with placental weights, placental diameters, placental coefficient and Capillarization index: with the increase of birth weights there is increase in these parameters.

C. HISTOLOGICAL EXAMINATION OF THE PLACENTAE

Histological examination of placentae of both control and IUGR specimens reveals the maternal component (decidual plate), the fetal component (chorionic plate) and the intervillous spaces. The decidual plate is formed of decidua basalis and placental septa which involve a core of maternal tissue covered by a layer of syncytiotrophoblasts. The chorionic plate comprises the chorionic villous tree (stem, intermediate and terminal villi) that is formed of a layer of trophoblasts surrounding the villous stroma and villous vessels. Some villi show clumps of syncytiotrophoblastic nuclei protruding from the villous surface forming syncytial knots. The intervillous spaces are filled with maternal blood (Fig. 3).

In the deciduas, trophoblast giant cells (TGCs) with large sized polypoid nuclei and prominent wide cytoplasm are present within the trophoblast cell layer and close to the decidual vessels (Fig. 4). In the chorionic villi, TGCs are noticed within the trophoblast cell layer, in the syncytial knots and in the nearby the chorionic vessels (Fig. 5).

D. PATHOLOGICAL EXAMINATION OF THE PLACENTAE

The stem villi of many term idiopathic IUGR placentae show hyalinization of the interstitium and hypercellular stromal response. Others show tiny or focal areas of villous necrosis and or fibrosis (Fig. 6). Many stem villous arteries of term idiopathic IUGR placentae show marked wall hypertrophy and narrowing of their lumina. Also, many stem villous arteries show wall herniation, hemorrhagic dissection of the vessel wall, focal inflammatory infiltrate and or thrombosis (Fig. 7). The terminal villi of some specimens show marked fibrotic changes (Fig. 6).

Varying numbers of control term placentae show some of the pathological findings exhibited by the term idiopathic IUGR placentae.

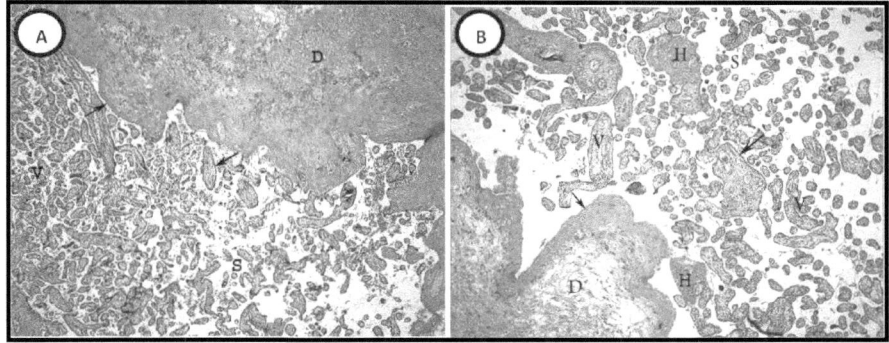

Fig. 3: Representative histological sections of paraffin embedded placental tissues of control (A) and IUGR (B) specimens reveal; decidua basalis (D), chorionic villous tree (stem, intermediate and terminal villi) (V) and intervillous spaces (S). Decidua basalis is formed of a layer of syncytiotrophoblasts (thin arrow) surrounding maternal tissue with decidual blood vessels. Chorionic villi are formed of a layer of trophoblasts (thick arrow) surrounding the villous stroma and villous vessels. Note the hyalinization in some villi of IUGR section (H) (HE, x100).

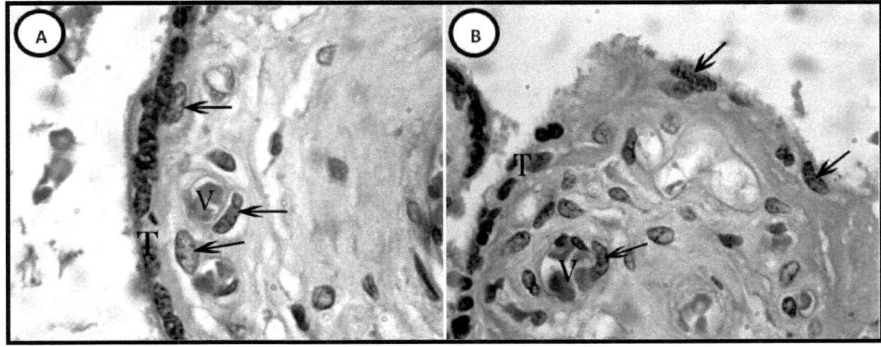

Fig. 4: Representative histological sections of paraffin embedded placental tissues; reveals the deciduas of control (A) and IUGR (B) specimens with their lining trophoblast cell layer (T) and decidual blood vessels (V). The trophoblastic giant cells with its large sized polypoid nuclei and prominent wide cytoplasm appear

within the trophoblastic cell layer and in the nearby the decidual vessels (arrows) (HE, x1000).

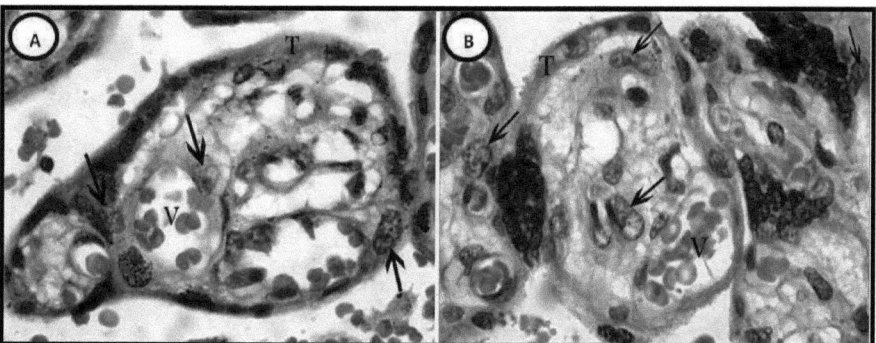

Fig. 5: Representative histological sections of paraffin embedded placental tissues; reveals the chorionic villi of control (A) and IUGR (B) specimens with their covering trophoblast cell layer (T) and villous vessels (V). The trophoblast giant cells with their large sized polypoid nuclei and prominent wide cytoplasm appear within the trophoblastic cell layer, in the syncytial knots and close to the villous vessels (arrows) (HE, x1000).

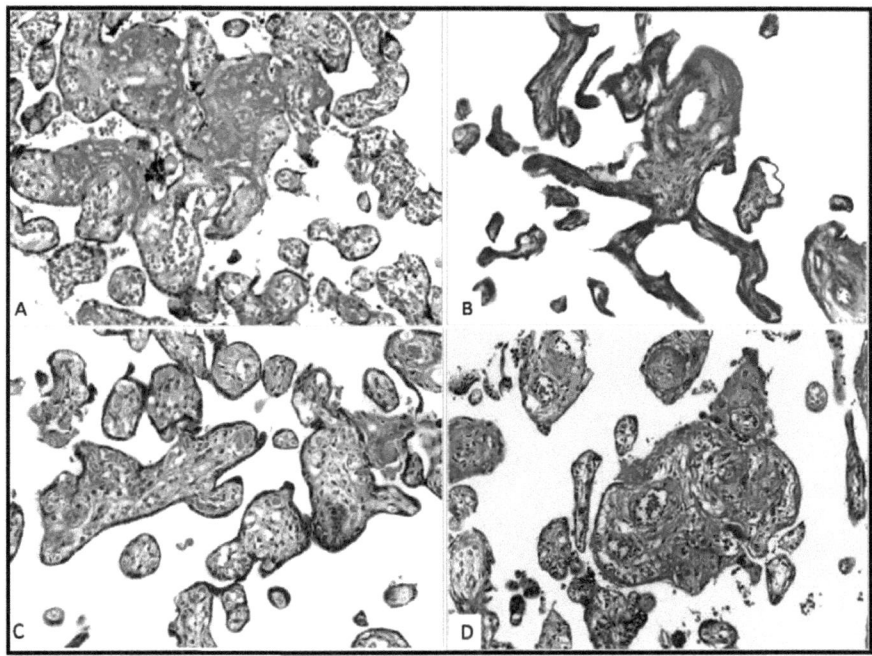

Fig. 6: Photomicrograph of a histological section of placenta from a pregnancy with idiopathic intrauterine growth restriction showing: (A) The intervillous space with stem villous hyalinization and disappearance of stem arteries (HE, ×100). (B) Intervillous space with marked fibrosis of the villi (HE, x100). (C) Intervillous space with foci of cellular infiltration in the stem villi (HE, x100). (D) Intervillous space with stem villous hyalinization, hypercellular stromal response and tiny or focal areas of villous necrosis (HE, x100).

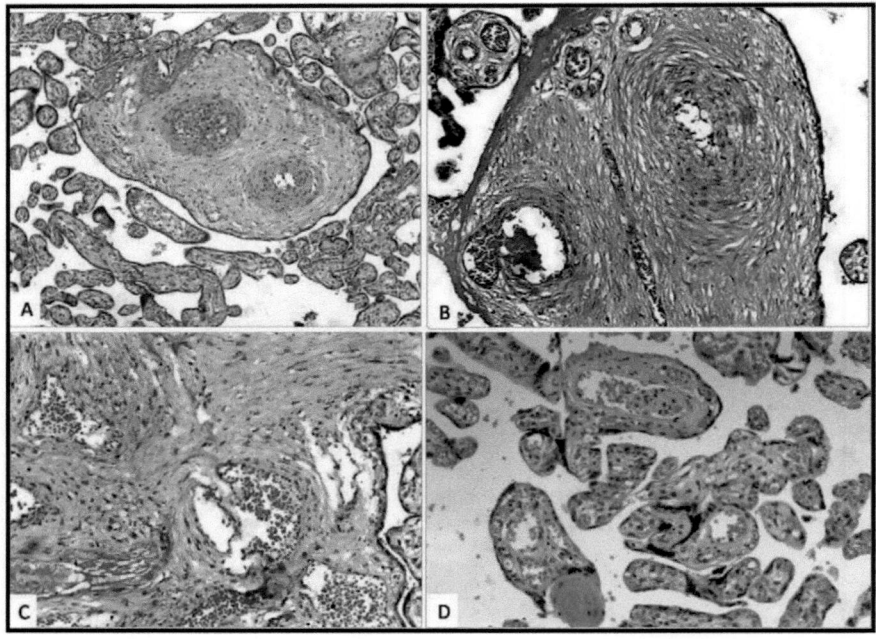

Fig. 7: Photomicrograph of a histological section of placenta from a pregnancy with idiopathic intrauterine growth restriction showing stem villi demonstrating: (A) stem artery with marked wall hypertrophy and narrowing of the lumen (hematoxylin and eosin, original magnification ×100). (B) stem villous artery with herniation of its wall, other artery appears with marked wall hypertrophy and narrowing of its lumen (hematoxylin and eosin, original magnification ×200). (C) Stem villous artery with hemorrhagic dissection of its wall. Hyalinization and cellular infiltration is apparent in the villous interstitium (hematoxylin and eosin, original magnification ×200). (D) Stem villous artery with intraluminal thrombus (HE, ×100).

Statistical analysis of the histopathological findings in placentae

- **The frequency of different pathological conditions as shown in Tables (7, 8):**

Image analyzing study of histological sections from term idiopathic IUGR show reduced number of placental villous stem arteries compared with control specimens. The difference between mean number of arteries per field in control (12.36 ± 0.61) and idiopathic IUGR (4.63 ± 0.46) group is statistically significant (p=0.000). Also, in idiopathic IUGR group the number of capillaries in the terminal villi per field is decreased compared with control group. The difference between mean number of capillaries per field in AGA (73.35 ± 5.13) and idiopathic IUGR (47.09 ± 4.44) group is statistically significant (p=0.000).

The degenerative changes (hyalinization and necrosis) in the stem villi and the presence of villitis are significantly higher in idiopathic IUGR cases than in control ones (p=0.000, p=0.001). Whereas, the degenerative or fibrotic changes in the terminal villi show no significant difference between both groups (p=0.370).

The narrowing of stem villous arteries is significantly higher in idiopathic IUGR than in control group (p= 0.001), whereas the degenerative findings in the stem arterial wall show no significant differences between both groups (p= 0.588).

- **The correlation between the main pathological findings in the placental tissue:**

The number of stem villous arteries is significantly correlated with the presence of villitis (r=0.243; p=0.013) and the degenerative changes in the stem villi (r=0.341; p=0.001). It is also correlated with the terminal capillary number (r=0.253; p=0.002) and the fibrous changes in the terminal villi (r=-0.243; p=0.018).

The narrowed lumen of stem arteries is significantly correlated with the presence of villitis in stem villi (r=0.324; p=0.005) and with the terminal capillary number (r=0.129; p=0.046).

Interestingly, we found that the fibrotic changes in the terminal villi are significantly correlated with the number of stem villous arteries (r=-0.243; p=0.018) but not significantly correlated with the number of capillaries in the terminal villi (r=-0.120; p=0.830).

- **Fetal and placental correlation with the pathological changes:**

The birth weight and the placental weight are significantly correlated with the appearance of degenerative changes in the stem villi (r=0.33; p=0.001, r=0.345; p=0.000).

The birth weight is significantly correlated with stem arteries number (r=0.494; p=0.000), arterial narrowing (r=0.283, p=0.004), terminal villous capillary number (r=0.281, p=0.001) and villitis (r=0.275, p=0.005).

The placental weight is significantly correlated with stem arteries number (r =0.494, p=0.000), arterial narrowing (r=0.283, p=0.004), villitis (r=0.252, p=0.009) and terminal villous capillary number (r=0.281, p=0.000).

The average placental diameter is significantly correlated with stem arteries number (r=0.330, p=0.000), arterial narrowing (r=0.23, p=0.023), villitis (r=0.301, p=0.002) and terminal villous capillary number (r=0.168, p=0.042).

Table 7: Mean number of stem villous arteries and terminal villous capillaries in control and idiopathic IUGR groups

Variable	Control group n=25 Mean±SEM	Idiopathic IUGR group n=50 Mean±SEM	P value
Number of stem villous arteries per field.	12.36 ± 0.61	4.63 ± 0.46	0.000*
Number of terminal villous capillaries per field	73.35 ± 5.13	47.09 ± 4.44	0.000*

Data are presented as mean ± standard error of means (SEM). Significance was taken at P<0.05 for Independent Samples-t test (*). IUGR: intrauterine growth restriction.

Table 8: Frequency of different pathological conditions in placentae of control and idiopathic IUGR groups

The pathological condition	Freqency in control group	Freqency in idiopathic IUGR group	P value
Narrowing of stem villous arteries	9 (36%)	38 (76%)	0.0 01*
Degeneration of the stem arterial wall (herniation, hemorrhagic dissection, focal inflammatory infiltrate and or thrombosis).	6 (24%)	15 (30%)	0.5 88
Hyalinization and necrosis of the stem villi	13(52%)	48 (96%)	0.0 00*
Cellular infiltration (villitis) of the stem villi	2 (8%)	24 (48%)	0.0 01*
Terminal villous degeneration or fibrosis	3 (12%)	3 (6%)	0.3 70

Significance was taken at P<0.05 for Mann–Whitney U-test (*).

Discussion

The present work studied the histomorphological and pathological changes in the stem and terminal villi of idiopathic IUGR and control placentae aiming to reach probable pathogenesis of idiopathic IUGR in Saudi. The stem villi of term idiopathic IUGR placentae showed different pathologic findings including arterial changes, degenerative changes and signs of villitis. Marked fibrotic changes are presented in the terminal villi of some specimens. Importantly, these pathologic pictures are also exhibited by varying numbers of control term placentae.

The pathologic findings seen in this study are in line with Salafia et al. (1992) who reported the presence of placental infarction, chronic villitis, hemorrhagic endovasculitis, and placental vascular thromboses in different cases of idiopathic IUGR at term. The presence of these pathologic findings in both IUGR and control cases is in parallel with Tomasa et al. (2010) who reported that there was no difference in histopathologic findings between idiopathic IUGR placentae and control ones. Also Salafia et al. (1992) found that one or more pathologic changes were present in 55% of IUGR cases, and 32% of non-IUGR cases.

In this study, the mean number of stem villous arteries, and the mean number of terminal villous capillaries per field are significantly lower in idiopathic IUGR (4.63 ± 0.46, 47.09 ± 4.44) than in control group (12.36 ± 0.61, 73.35 ± 5.13 respectively) (p=0.000, p=0.001 respectively). In line with our results, Giles et al. (1985) and Sebire (2003) suggested that placentae from pregnancies with IUGR and abnormal umbilical artery (UA) Doppler findings were associated with a reduced number of placental villous stem arteries. Subsequently, other authors appeared to confirm these findings (McCowan et al. 1987; Bracero et al. 1989). Meanwhile, many studies with systematic sampling techniques have been unable to confirm such results (Hitschold et al. 1993; Jackson et al. 1995). In controversy with our results, Claude and Steven (1985) and Lena et al. (1995) reported increased mean number of capillaries in the placental tissue and explained that the increased number of capillaries in the stromal core of terminal villi indicates that hypoxia induces hypercapillarization and vasodilatation in many systemic vessels, but causes vasoconstriction in pulmonary vessels.

In the present work, narrowing of stem villous arteries is significantly higher in idiopathic IUGR (76%) than in control (36%) cases. This apparent narrowing might be due to wall hypertrophy and or vasoconstriction. Several authors have reported the apparent luminal reduction and wall hypertrophy of stem vessels in placentae from cases of IUGR consistent with marked, longstanding placental stem villous vasoconstriction (Van der Veen and Fox 1983; Sebire et al. 2001). Many authors hypothesized that intrauterine growth restriction, with or without etiology leads to chronic stress to the fetus with chronic hypoxia and release of vasoactive substances, which cause chronic vasoconstriction and vascular hypertrophy (Subahash et al. 2000; Cunningham et al. 2010).

On the other hand, our results proved that the vascular degenerative changes of the stem arteries (wall herniation, hemorrhagic dissection of the vessel wall, focal inflammatory infiltrate and or thrombosis) showed higher but not significant difference in idiopathic IUGR (30%) than in control (24%) cases. In comparison with the present results, Salafia et al. (1992) found that hemorrhagic endovasculitis was exhibited in 15% of all IUGR cases, whereas placental vascular thromboses presented in 9% of cases.

Our results showed that, the presence of villitis and the degenerative changes in the stem villi are significantly higher in idiopathic IUGR cases than in control ones. In accordance with these results, (Salafia et al. 1992) reported higher percentage of chronic villitis and placental infarction in idiopathic IUGR cases (53%, 63% respectively) than in control cases.

Discussing the causative mechanisms of idiopathic IUGR, the present study proved significant positive correlation between the birth weight and different pathologic features in the stem villi as stem artery number (r=0.494; p=0.000), arterial narrowing (r=0.283, p=0.004), stem villous degenerative changes (r=0.331, p=0.001) and villitis (r=0.275, p=0.005) and also significant positive correlation between birth weight and the terminal villous capillary number (r=0.281, p=0.001) but no significant correlation between birth weight and the terminal villous fibrotic changes (r=-0.098, p=0.318). These results could raise the hypothesis that the stem villi represent the mystery for the development of idiopathic IUGR.

In accordance with our finding that decreased number of stem arteries and or their narrowing might be causative mechanisms for decreased birth weight and development of idiopathic IUGR in Saudi. Sebire (2003) hypothesized that reduced placental stem artery number could be the first mechanism proposed in IUGR and abnormal UA Doppler findings. Also, Campbell et al. (1986) and Bower et al. (1991) reported the reduction in uteroplacental blood flow as an underlying event in most cases of IUGR. In the same time, Fox (1997) found that the villous ischemic necrosis associated with severe IUGR is secondary to severe reduction in oxygen delivery due to severe localized impairment of uteroplacental intervillous blood flow. This is in accordance with the pathological finding of defective endovascular trophoblast invasion and conversion of uteroplacental vessels in IUGR cases (Brosens 1977).

That the stem arterial narrowing could be an underlying mechanism in idiopathic IUGR is in line with Sebire et al. (2001) and Sebire and Talbert (2002) who suggested the placental stem villous vasoconstriction as underlying mechanism in IUGR with abnormal UA Doppler waveforms and explained the anatomical importance of the vascular smooth muscle of placental stem villous arteries and veins which are well developed suggesting their important role in controlling placental hemodynamic to minimize 'ventilation–perfusion mismatch. Histopathological evidence for such changes compatible with prolonged vasoconstriction have now been well described in stem vessels from placentae in pregnancies with IUGR by several groups (Van der Veen and Fox 1983; Sebire et al. 2001) and the current theory is that prolonged reduction in maternal intervillous flow, and hence oxygen delivery, would therefore result in prolonged vasoconstriction with secondary reduction in stem villous luminal diameter, increased flow resistance and vascular medial hypertrophy. The precise mechanism of control of stem villous vessel tone has not previously been determined with certainty, although release of vasoactive mediators from the distal terminal villi, such as nitric oxide (NO), has been suggested (Myatt et al. 1991; Hampl et al. 2002).

The present study demonstrated that birth weight is significantly correlated with cellular infiltration and or degenerative changes in the stem villi. This could raise the hypothesis that stem villous degeneration and or villitis especially of unknown etiology (VUE) could be underlying mechanisms of idiopathic IUGR. In accordance with this hypothesis, Raymond and Redline (2007) stated that VUE (When low-grade lesions are excluded) is an important cause of intrauterine growth restriction and recurrent reproductive loss. Also, Salafia et al. (1992) proved that 30% of all idiopathic IUGR cases were having chronic villitis that could be accompanied by hemorrhagic endovasculitis.

That the terminal villous mal-development could be a probable causative mechanism of idiopathic IUGR is a point of discrepancy. Our study proved that the birth weight is positively correlated with terminal villous capillary number (r=0.281, p=0.001) but not with the terminal villous fibrotic changes (r=-0.098, p=0.318). The number of terminal villous capillaries is also a point of debate where many researchers prove hypocapillarization of terminal villi with IUGR (Mayhew et al. 1999; Mayhew et al. 2004; Egbor et al. 2006), while others prove hypercapillarization (Claude and Steven 1985; Lena et al. 1995). On the other hand, pathological studies of placentae with IUGR and abnormal UA Doppler findings reported that the terminal villi are often small, hypovascular and fibrotic which prompted the hypothesis that primary villous maldevelopment may be the underlying event in such cases (Macara et al. 1995; Macara et al. 1996). In consistent with our results, Sebire (2003) declared that the underlying mechanism of abnormal UA Doppler waveforms in most cases of IUGR is likely to be secondary to significant reduction in maternal uteroplacental flow rather than primary placental/villous maldevelopment and described that resistance to flow in almost all organs is controlled at the level of the small arteries/arterioles, not at the level of the capillary bed itself. This could be in consistent with our findings that the fibrotic changes in the terminal villi are significantly correlated with the number of stem villous arteries (r= -0.243; P= 0.018) but not with the number of capillaries in the terminal villi (r= -0.120; P= 0.830).

In conclusion, histomorphological and pathological changes in the stem villi could explore the cause of idiopathic IUGR. The decreased number of stem arteries, narrowing of the arterial walls, degeneration of stem villi and villitis could be underlying mechanisms. Further researches on the hormonal and cytokine level should be undertaken to demonstrate the precipitating factors of these changes and the possible preventing measures.

E. TNF-α IMMUNOHISTOCHEMICAL ANALYSIS

TNF-α immunostaining is detected in both control and idiopathic IUGR specimens. In the deciduas, TNF-α is localized in the cell membranes and cytoplasm of decidual trophoblast and in the endothelium of decidual vessels. TGCs in the deciduas of control specimens show deficient or negative TNF-α immunoexpression while those of IUGR group show positive staining (Fig. 8). In the chorionic villi, TNF-α is expressed in the cell membranes and cytoplasm of the villous trophoblasts, endothelium of chorionic vessels and in the syncytial knots. Also, TGCs of the chorionic plates of control specimens show deficient or negative TNF-α immunoexpression while those of IUGR group that show positive staining (Fig. 9).

Statistical study of the image analysis data of all specimens reveal that the mean area percent of TNF-α immunostainaing is significantly higher in the idiopathic IUGR group (5.93±0.69) compared to the control one (3.28±0.41) (p=0.001) (Figs. 8, 9).

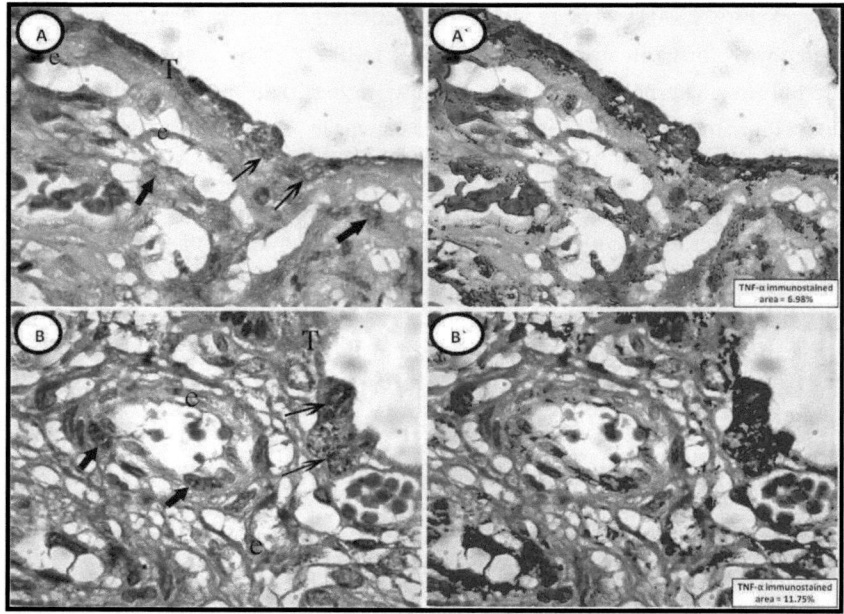

Fig. 8: Representative TNF-α immunostaining of paraffin embedded placental tissues; reveals the localization and area percent of TNF-α in the deciduas of control (A, A`) and IUGR (B, B`) specimens: In both deciduas, TNF-α is localized in the cell membranes and cytoplasm of decidual trophoblast (T) and the endothelium of the decidual vessels (e) with higher area percent in FGR specimens. The trophoblast Giant cells located within the trophoblastic cell layer (thin arrows) or close to the spiral artery (thick arrow) show negative TNF-α immunostaining in control chorions (A, A`) but positive immunostaining in IUGR (B, B`) ones (anti-TNF-α, x1000).

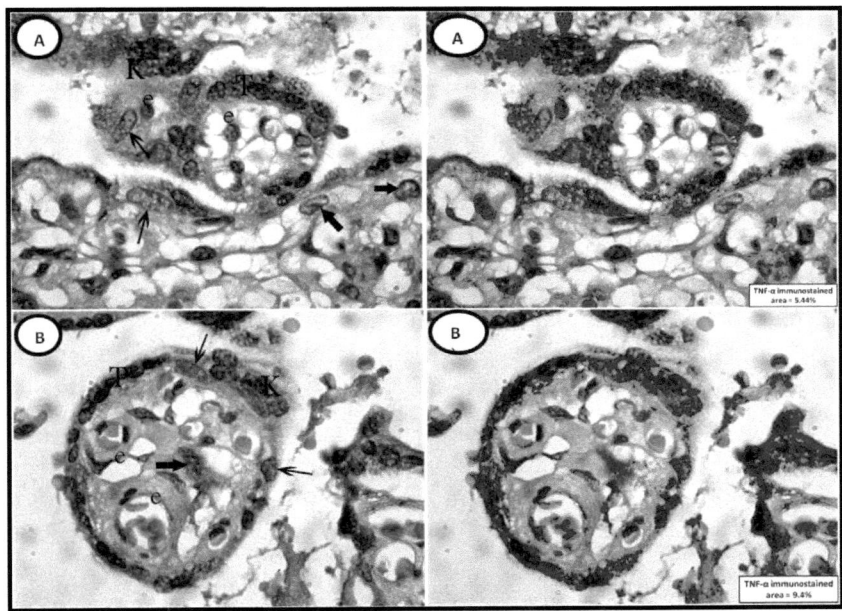

Fig. 9: Representative TNF-α immunostaining of paraffin embedded placental tissues; reveals the localization and area percent of TNF-α in the chorionic villi of control (A, A`) and IUGR (B, B`) cases: In both chorions, TNF-α is localized in the cell membranes and cytoplasm of the villous trophoblast (T), syncytial knots (K) and in the endothelium of villous vessel walls (e) with higher area percent in IUGR specimens. The trophoblast giant cells located within the trophoblastic cell layer (thin arrows), and that associated with the villous vessels (thick arrows) show negative TNF-α immunostaining in control chorion (A, A`) but positive immunostaining in IUGR (B, B`) specimens (anti-TNF-α, x1000).

Discussion

Considering the fact that growth restriction is a major risk factor for newborns, this study assesses the expression and the localization of TNF-α in the placental tissue in cases of idiopathic IUGR hoping to clarify the associated histological and immunohistochemical changes. In the present work, TNF-α immunoexpression is detected in the all examined full-term placental tissues (decidual and chorionic plates) whether from control or idiopathic IUGR pregnancies, with significantly higher area percent in the idiopathic IUGR group.

In this study, the deciduas of both control and idiopathic IUGR placentae exhibit TNF-α immunoexpression in the cell membranes and cytoplasm of the decidual trophoblast, and in the endothelium of the blood vessels. As well, in the chorionic villi, it is localized in the cell membranes and cytoplasm of the villous trophoblasts, endothelium of blood vessels and in the syncytial knots. This expression of TNF-α in full-term control placentae, is matched with the finding of Wang et al. (2008) who demonstrated TNF-α in normal human placental tissues signifying its changes during pregnancy as an indicative of its specific function in the developmental differentiation processes. Expression of TNF-α might change regarding to the needs of development; early in gestation, TNF mRNA seems to be predominantly expressed in all cell types of the trophoblastic lineage (Yang et al. 1993; King et al. 1995). Then, as pregnancy proceeds, mRNA expression switches from the trophoblastic cell population to a stronger signal within villous stromal cells (Hung et al. 2004). In situ hybridization and immunohistochemical staining studies had localized the expression of TNF receptors to trophoblasts of the placental villi (Yelavarthi and Hunt 1993).

The observation that TNF-α induces apoptosis of primary human trophoblasts (Yui et al. 1994; Garcia-Lloret et al. 1996) suggests both physiological (trophoblast turnover) and pathological (loss of the protective trophoblast barrier) roles in the placenta. According to the present results, the mean percentage of TNF-α immunostained areas is significantly higher in idiopathic IUGR placental tissues compared to control placentae suggesting a specific role of TNF-α in the development of IUGR and might explain the histopathological theories for development of these cases. In consistent with this assumption, Hunt et al. (1990) and Wride and Sanders (1995) stated that TNF-α may inhibit the growth of the trophoblasts and it can control the programmed cell death and remodeling of the extracellular matrix. From other point of view, the increased TNF-α expression in restricted fetal growth could be a priming event for the trophoblasts to undergo inflammation process as was stated by Nawroth and Stern (1986).

In the blood vessels, TNF-α is detected in the vascular endothelial cells of the decidual and villous vessels which might reflect the cause of growth restriction owing to failure of trophoblast invasion and underperfusion of the uteroplacental bed as stated by Lyall et al. (1999). Further, TNF-α is directly toxic to endothelium and may damage the decidual vasculature (Hunt et al. 1990). Still, it interferes with the anticoagulant system and may induce placental thrombosis (Bevilacqua et al. 1986). Recently, and in consistent with the present results, Holcberg et al. (2001) proved the increased TNF secretion in placentae of IUGR fetuses and attributed this to the enhanced vasoconstriction of the fetal placental vascular bed. Fortunately, the increased TNF-α tissue expression in idiopathic IUGR proved in this study, is in consistent with other studies reporting increased maternal serum and amniotic fluid levels of this cytokine in IUGR (Stallmach et al. 1995; Holcberg et al. 2001).

TGCs are the terminally differentiated, multinucleated, invasive cells of the developing placenta, responsible for remodeling the extracellular matrix of the uterine stroma as implantation progresses. The present study demonstrated the histological characteristics of TGCs comprising a large sized polypoid nuclei and prominent wide cytoplasm matching with that previously described by Bevilacqua and Abrahamsohn (1988). In the present study, the full term placental tissue of control and idiopathic IUGR pregnancies, reveal the presence of TGCs within the trophoblastic cell layer and close to the vessels of both decidua and chorionic villi which is in consistent with Simmons et al. (2007).

The results of the present work reveal positive TNF-α expression in TGCs located within the decidua and chorion of the IUGR placentae but deficient or negative TNF-α expression in those of control placentae. This could assume TGC as one of the cell-sources for increased level of this cytokine in idiopathic IUGR. Some studies have revealed the TGCs of the early placenta as the site of TNF-α expression (Hunt et al. 1993; Lachapelle et al. 1993) but little studies have demonstrated TNF-α in TGCs in full-term placental tissues. In accordance with our results, Pijnenborg et al. (1998) proved that at later stages of normal pregnancy TNF protein expression decreases in invasive cells and that the trophoblast giant cells lack any TNF expression. However, no literature is found to discuss the expression of TNF-α in TGCs of idiopathic IUGR placentae.

In conclusion, this study has provided a comprehensive expression profile for TNF-α in normal and idiopathic IUGR full-term placental tissue, demonstrating enhanced TNF-α immunoexpression in idiopathic IUGR placentae. The study also has proposed the trophoblastic giant cells to be an obvious source of this cytokine in such cases. Experimental studies should be done to ascertain the effect of increased TNF-α as inflammatory cytokine on the fetal development and the possible opposing effects of anti-inflammatory cytokines.

F. TdT IMMUNOHISTOCHEMICAL ANALYSIS

The TdT-immunostaining in villous tissue varied in the studied groups. In control placental villi, TdT- immunostaining was present as score 1+ (less than 50% staining) in the syncytiotrophoblasts and as score 0 (no staining) in the cytotrophoblasts and stroma cells. Compared with control group, the intensity of TdT-immunostaining in idiopathic IUGR is detected as score 2+ (50% or more staining) in the syncytiotrophoblasts and as score 1+ in the cytotrophoblasts and stroma cells with significant difference between both groups (p<0.005) (Fig. 10).

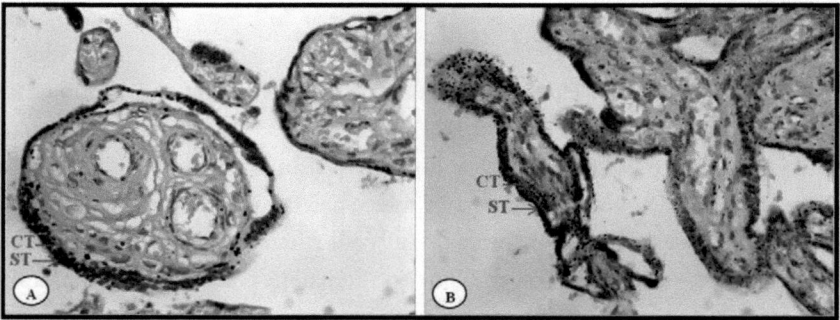

Fig. 10: Representative TdT-immunostaining of paraffin embedded placental tissues showing: (A) chorionic villi of control placenta with score 1+ nuclear staining for TdT in the syncytiotrophoblasts and score 0 staining in the cytotrophoblasts and stroma cells. (B): chorionic villi of idiopathic IUGR placenta with score 2+ nuclear staining for TdT in the syncytiotrophoblasts and score 1+ staining in the cytotrophoblasts and stroma cells. (anti-TdT, x400).

Discussion

In placentae, apoptosis is involved in normal development, differentiation, remodeling and in their aging process (Austgulen et al. 2002). The present study evaluates the incidence of apoptosis in the placentae of control and idiopathic IUGR pregnancies. In both groups, apoptosis is highly represented in syncytiotrophoblasts than in cytotrophoblasts and stroma cells. This might indicate that syncytiotrophoblasts are more prone to programmed cell death than their progenitor cells (cytotrophoblasts).

This study shows that the rate of apoptosis is significantly higher among the cells of placentae of pregnancies complicated with idiopathic IUGR than the cells of the placentae of normal uncomplicated pregnancies. These results are matched with Smith et al. (1997); Axt et al. (1999) who clearly demonstrated the distribution of apoptotic bodies in placental tissues of pregnancies complicated with IUGR. However, Kokawa et al. (1998); Smith and Baker (1999) proved the increased rate of apoptosis in the placental tissue from pregnancies complicated with other cases than IUGR.

The most important question raised by this study is whether the increased rate of apoptosis in the placental tissue is a result of an etiopathologic factor leading to IUGR or a compensatory mechanism for the nutritional transport and gas exchange to meet the metabolic requirements of the fetus. As an etiopathologic factor, hypoxia is a known trigger of apoptosis in different tissues including the placenta (Graeber et al. 1996). This could explain our results that a higher rate of apoptosis appears in placentae of pregnancies complicated with IUGR as the placenta may respond to hypoxic stress by enhancing the number of apoptotic cells.

There has been controversy about the localization of apoptosis in the placental tissue. In this study, and in support with other studies (Smith et al. 1997; Nelson 1996), most of the apoptotic cells are presented within the trophoblast, particularly the syncytiotrophoblasts. This could be due to the fact that, in the third trimester syncytiotrophoblast is more prevalent and in turn, the rate of apoptosis should be expected to be higher. Also, in accordance with our results, Smith et al. (1997) proved that apoptosis is more prevalent in syncytiotrophoblast than cytotrophoblast in the placentae of pregnancies complicated with IUGR. Nevertheless, in opposition to the present study, Kokawa et al. (1998) demonstrated the predominance of apoptosis in cytotrophoblast in samples of normal placenta and DiFederico et al. (1999) proved a widespread apoptosis of cytotrophoblasts in the placentae of pregnancies complicated with pre-eclampsia. Moreover, in the same cases, decreased cytotrophoblast proliferation was reported (Fitzgerald et al., 2011).

The increased apoptosis in the trophoblast in this study might be due to the increased secretion of TNF-α or other cytokines stimulating apoptosis by trophoblastic cells as was proved by Hunt et al. (1992) and Huppertz et al. (1999) in cases of preeclampsia.

In conclusion, the rate of apoptosis is increased in the placentae of pregnancies complicated with IUGR. Apoptosis is more abundant in the trophoblasts and it may have an important role in the pathogenesis of placentae of pregnancies complicated with IUGR.

REFERENCES

1. Akturk A, Onal ES, Atalay Y, Yurekli M, Erbas D, Okumus N, Turkyilmaz C, Unal S, Ergenekon E, Koc E, Himmetoglu O. Maternal and umbilical venous adrenomedullin and nitric oxide levels in intrauterine growth restriction. J of Maternal–fetal and Neonatal Medicine 2007;20(7):521-5.

2. Austgulen R, Chedwick L, Vogt Isaksen C, Vatten L, Craven C (2002): Trophoblast apoptosis in human placenta at term as detected by expression of a cytokeratin 18 degradation product of caspase. Arch Pathol Lab Med, 126:1480–6.

3. Axt R, Kordina AC, Meyber R, Reitnauer K, Mink D, Schmidt W (1999): Immunohistochemical evaluation of apoptosis in placentae from normal and intrauterine growth-restricted pregnancies. Clin Exp Obstet Gynecol 26:195-198.

4. Bassett JM, Hanson C. Catecholamines inhibit growth in fetal sheep in the absence of hypoxemia. American Journal of Physiology. 1998;274:R1536–1545.

5. Bevilacqua EM, Abrahamsohn PA (1988) Ultrastructure of trophoblast giant cell transformation during the invasive stage of implantation of the mouse embryo. J Morphol. 198:341-51.

6. Biswas S. and Ghosh S.K. (2008): Gross morphological changes of placentas associated with intrauterine growth restriction of fetuses: A case control study. Early Human Development. 84, 357–362.

7. Bocci G, Fasciani A, Danesi R, Viacava P, Genazzani AR, Del Tacca M (2001): In-vitro evidence of autocrine secretion of vascular endothelial growth factor by endothelial cells from human placental blood vessels. Mol. Hum. Reprod. 7:771–7.

8. Bower S, Schuchter K, Campbell S (1991) Doppler ultrasound screening as part of routine antenatal scanning: prediction of pre-eclampsia and intrauterine growth retardation. Br J Obstet Gynaecol 98: 871–879.

9. Boyle DW, Lecklitner S and Liechty EA (1996) Effect of prolonged uterine blood flow reduction on fetal growth in sheep American Journal of Physiology 270 R246–R253

10. Bracero LA, Beneck D, Kirshenbaum N, Peiffer M, Stalter P, Schulman H (1989) Doppler velocimetry and placental disease. Am J Obstet Gynecol 161:388–392.

11.Briana DD, Malamitsi-Puchner A (2009): Intrauterine growth restriction and adult disease: the role of adipocytokines. Eur J Endocrinol. 160(3):337-47.

12.Brosens IA (1977) Morphological changes in the utero-placental bed in pregnancy hypertension. Clin Obstet Gynecol 4:573–593.

13.Campbell S, Pearce JM, Hackett G, Cohen-Overbeek T, Hernandez C. Qualitative assessment of uteroplacental blood flow: early screening for test for high-risk pregnancies (1986) Obstet Gynecol 68:649–653.

14.Chen G, Wilson R, Wang SH, Zheng HZ, Walker JJ, McKillop JH. TNF-α gene polymorphism and expression in preeclampsia. Clin Exp Immunol. 1996; 104: 154-9.

15.Chen HL, Yang Y, Hu XL, Yelavarthi KK, Fishback JL and Hunt JS (1991): Tumor necrosis factor alpha mRNA and protein are present in human placental and uterine cells at early and late stages of gestation. Am. J. Pathol. 139, 327–335.

16.Cheung BM, Lau CS, Leung RY, Tong KK, Kumana CR. Plasma adrenomedullin level in systemic lupus erythematosus. Rheumatology 2000;39:804–5.

17.Choi HK, Choi BC, Lee SH, Kim JW, Cha KY, Baek KH (2003): Expression of angiogenesis- and apoptosis-related genes in chorionic villi derived from recurrent pregnancy loss patients.

18.Claude G, Steven GS (1985) Pathology in gynaecology and obstetrics. Philadelphia : JB Lippincott Company 3:448-97.

19.Cormier-Regard S., Nguyen S.V., Claycomb W.C. (1998) Adrenomedullin gene expression is developmentally regulated and induced by hypoxia in rat ventricular cardiac myocytes.J. Biol. Chem. 273 17787–17792.

20.Cunningham FG, Leveno KJ, Bloom SL, Hauth JC, Rouse DJ, Spong CY. (2010). Disorders of amniotic fluid volume. In: Williams Obstetrics (23rd edition, pp.490-499). New York: McGraw-Hill Medical Publishing Division.

21.De Onis M, BlÖssner M, Villar J. (1998): Levels and patterns of intrauterine growth retardation in developing countries. Eur J Clin Nutr ;52:S5-15.

22.Di Iorio R, Marinoni E, Letizia C, Gazzolo D, Lucchini C, Cosmi EV (2000) Adrenomedullin is increased in the fetoplacental circulation in intrauterine growth restriction with abnormal umbilical artery waveforms. Am J Obstet Gynecol;182:650–4.

23.Di Iorio R, Marinoni E, Letizia C, Cosmi EV (2003) Adrenomedullin in prenatal medicine. Regul Pept;112:103-113.

24.DiFederico E, Genbacev O, Fisher SJ (1999): Preeclampsia is associated with widespread apoptosis of placental cytotrophoblasts within the uterine wall. Am J Pathol155:293-301.

25.Donald W, Li MH, Dakour J, Kaufman S, Guilbert LJ, Lowen BW (2003): Adrenomedullin Is Decreased in Preeclampsia Because of Failed Response to Epidermal Growth Factor and Impaired Syncytialization, Hypertension, 42;895-900.

26.Dudley DJ, Trautman MS, Araneo BA, Edwin SS & Michell MD. (1992): Decidual cell biosynthesis of interleukin-6: regulation by inflammatory cytokines. Journal of Clinical Endocrinology and Metabolism, 74 884–889.

27.Edmund R, Novak, Woodruff J, Donald (1979) Gynecologic and obstetric Pathology, 8th ed. London : W.B Saunder. 585-627.

28.Egbor M, T. Ansari T, Morris N, Green CJ and Sibbons PD (2006) Pre-eclampsia and Fetal Growth Restriction: How Morphometrically Different is the Placenta? Placenta 27, 727e734.

29.Fitzgerald B, Levytska K, Kingdom J, Walker M, Baczyk D, Keating S (2011): Villous trophoblast abnormalities in Extremely pretermDeliveries with elevated second trimester maternal serum hCG or Inhibin-A. Placenta. 32(4):339-45.

30.Fox H. Macroscopic abnormalities of the placenta (1997) In Pathology of the Placenta, FoxH (ed.). W. B. Saunders: London, UK 102–150.

31.Gagnon R, Murotsuki J, Challis JR, Fraher L, Richardson BS. Fetal sheep endocrine responses to sustained hypoxemic stress after chronic fetal placental embolization. American Journal of Physiology. 1997;272:E817–823.

32.Garcia-Lloret MI, Yui J, Winkler-Lowen B, Guilbert LJ (1996) Epidermal growth factor inhibits cytokine-induced apoptosis of primary human trophoblasts. J Cell Physiol. 167:324-332.

33.Giles WB, Trudinger BJ, Baird PJ (1985) Fetal umbilical artery flow velocity wavelengths and placental resistance: pathological correlation. Br J Obstet Gynaecol 92:31-38.

34.Gill RM and Hunt JS (2004): Soluble receptor (DcR3) and cellular inhibitor of apoptosis-2 (cIAP-2) protect human cytotrophoblast cells controlinst LIGHT-mediated apoptosis. Am. J. Pathol. 165, 309–317.

35. Gill RM, Ni J and Hunt JS (2002): Differential expression of LIGHT and its receptors in human placental villi and amniochorion membranes. Am. J. Pathol. 161, 2011–2017.

36. Graeber TG, Osmanian C, Jacks T, Housman DE, Koch CJ, Lowe SW et al. (1996): Hypoxia-mediated selection of cells with diminished apoptotic potential in solid tumors. ature379:88-91.

37. Grimble RF. Inflammatory status and insulin resistance. (2002): Current Opinion in Clinical Nutrition and Metabolic Care 5:551–559.

38. Hadlock FP, Shah YP, Kanon DJ, Lindsey JV (1992) Fetal crown-rump length: reevaluation of relation to menstrual age (5e18 weeks) with high resolution real-time US. Radiology. 182:501-5.

39. Haider S, KnÖfler M (2009) Human tumour necrosis factor: physiological and pathological roles in placenta and endometrium. Placenta. 30:111-123

40. Hailman et al., 1996: Hailman, E., Albers, J.J., Wolfbauer, G., Tu, A.Y., and Wright, S.D., 1996, "Neutralisation and transfer of lipopolysaccharide by phospholipid transfer protein. " J Biol Chem. S.12172-12178.

41. Hampl V, Bibova J, Stranak Z, Wu X, Michelakis ED, Hashimoto K, Archer SL (2002) Hypoxic fetoplacental vasoconstriction in humans is mediated by potassium channel inhibition. Am J Physiol Heart Circ Physiol 283:H2440-H2449.

42. Heyborne KD, Witkin SS, McGregor JA. Tumor necrosis factor alpha in midtrimester amniotic fluid is associated with impaired intrauterine fetal growth. Am J Obstet Gynecol 1992;167:920–5.

43. Hinson JP, Kapas S, Smith DM. (2000): Adrenomedullin, a multifunctional regulatory peptide. Endocr Rev.; 21:138–167.

44. Hitschold T, Weiss E, Beck T, Hunterfering H, Berle P (1993) Low target birth weight or growth retardation? Umbilical Doppler flow velocity waveforms and histometric analysis of the fetoplacental vascular tree. Am J Obstet Gynecol 168:1260-1264.

45. Holcberg G, Huleihel M, Sapir O, Katz M, Tsadkin M, Furman B, Mazor M, and Myatt L. (2001): "Increased production of tumor necrosis factor-alpha by IUGR human placentae." European Journal of Obstetrics and Gynecology. Reproductive Biology. 94:69-72.

46.Hung TH, Charnock-Jones DS, Skepper JN, Burton GJ (2004) Secretion of tumor necrosis factor-alpha from human placental tissues induced by hypoxiareoxygenation causes endothelial cell activation in vitro: a potential mediator of the inflammatory response in preeclampsia. Am J Pathol. 164:1049-6.

47.Hunt J.S, Atherton R.A, Pace J.L. (1990): "Differential responses of rat trophoblast cells and embryonic fibroblasts to cytokines that regulate proliferation and class 1 MHC antigen expression." Journal of Immunology 145:184-191,.

48.Hunt JS, Chen H, Hu X, Pollard JW (1993) Normal distribution of tumor necrosis factor-a messenger ribonucleic acid and protein in the uteri, placentas, and embryos of osteopetrotic (op/op) mice lacking colony stimulating factor-1. Biol Reprod. 49:441-452

49.Hunt JS, Chen HL and Miller L (1996): Tumor necrosis factors: pivotal components of pregnancy? Biol. Reprod. 54, 554–562.

50.Hunt JS, Chen HL, Hu XL, Tabibzadeh S (1992) Tumor necrosis factor-alpha messenger ribonucleic acid and protein in human endometrium. Biol Reprod. 47:141-7.

51.Huppertz B, Frank HG, Kaufmann P (1999): The apoptosis cascade-morphological and immunohistochemical methods for its visualization. Anat Embryol (Berl);200:1–18.

52.Huppertz B, Frank HG, Kingdom JC, Reister F, Kaufmann P (1998): Villous cytotrophoblast regulation of the syncytial apoptotic cascade in the human placenta. Histochem Cell Biol, 110:495–508.

53.Jackson MR, Walsh AJ, Morroe RJ, Mullen JB, Lye SJ, Ritchie JW (1995) Reduced placental villous tree elaboration in small for gestational age pregnancies: relationship with umbilical artery Doppler waveforms. Am J Obstet Gynecol 172:518-525.

54.Johnson MR, N. Anim-Nyame, P. Johnson, S. R. Sooranna, and P. J. Steer, "Does endothelial cell activation occur with intrauterine growth restriction?" British Journal of Obstetrics and Gynecology, vol. 109, no. 7, pp. 836–839, 2002.

55.Jougasaki M, Burnett JC Jr. Adrenomedullin: Potential in physiology and pathophysiology. Life Sci 2000;66:855–72.

56.Kiernan JA (1999) Histological and Histochemical methods: Theory and practice. 3ed. Butterworth-Heinemann: Oxford.

57.King A, Jokhi PP, Smith SK, Sharkey AM, Loke YW (1995) Screening for cytokine mRNA in human villous and extravillous trophoblasts using the reversetranscriptase polymerase chain reaction (RT-PCR). Cytokine. 7:364-71.

58.Kokawa K, Shikone T, Nakano R (1998): Apoptosis in human chorionic villi and decidua during normal embryonic development and spontaneous abortion in the first trimester. Placenta 19:21-6.

59.Lachapelle MH, Miron P, Hemmings R, Falcone T, Granger L, Bourque J, Langlais J (1993) Embryonic resistance to tumor necrosis factor-a mediated cytotoxicity: novel mechanism underlying maternal immunological tolerance to the fetal allograft. Hum Reprod. 7:1032-1038.

60.Lena M, John CPK, Gaby K, Adrian W, Bowman, Peter K.(1995) Elaboration of stem villous vessels in growth restricted pregnancies with abnormal umbilical artery Doppler Waveforms. Br J of Obstet Gynaecol. 102:807-12.

61.Li H, Dakour J, Kaufman S, Guilbert LJ, Winkler-Lowen B, Morrish DW. Adrenomedullin Is Decreased in Preeclampsia Because of Failed Response to Epidermal Growth Factor and Impaired Syncytialization. Hypertension 2003;42:895.

62.Lyall F, Bulmer JN, Kelly H, Duffie E, Robson SC (1999) Human trophoblast invasion and spiral artery transformation: the role of nitric oxide. Am J Pathol. 154:1105-1114.

63.Macara L, Kingdom JC, Kaufmann P, Kohnen G, Hair J, More IA, Lyall F, Greer IA (1996) Structural analysis of placental terminal villi from growth-restricted pregnancies with abnormal umbilical artery Doppler waveforms. Placenta 17:37-48.

64.Macara L, Kingdom JCP, Kohnen G, Bowman AW, Greer IA, Kaufman P (1995) Elaboration of stem villous vessels in growth restricted pregnancies with abnormal umbilical artery Doppler waveforms. Br J Obstet Gynaecol 102:807-812.

65.Marinoni E, Pacioni K, Sambuchini A, Moscarini M, Letizia C, Di Iorio R (2011): Regulation by hypoxia of adrenomedullin output and expression in human trophoblast cells. European Journal of Obstetrics & Gynecology and Reproductive Biology;154: 146–150

66.Mayhew TM, Ohadike C, Baker PN, Crocker IP, Mitchell C, Ong SS (1999) Stereological investigation of placental morphology in pregnancies complicated by pre-eclampsia with and without intrauterine growth restriction. Placenta 24:219e26.

67.Mayhew TM, Wijesekara J, Baker PN, Ong SS (2004) Morphometric evidence that villous development and fetoplacental angiogenesis are compromised by intrauterine growth restriction but not by preeclampsia. Placenta 25:829-33.

68.Mayhew, T.M., 2002. Fetoplacental angiogenesis during gestation is biphasic, longitudinal and occurs by proliferation and remodeling of vascular endothelial cells. Placenta. 23, 742-50.

69.Mayhew, T.M., Sorensen, F.B., Klebe, J.G., Jackson, M.R., 1994. Growth and maturation of villi in placentae from well-controlled diabetic women. Placenta. 15, 57-65.

70.McCowan LM, Mullen BM, Ritchie K (1987) Umbilical artery flow velocity waveforms and the placental vascular bed. Am J Obstet Gynecol 157:900-902.

71.Meekins JW, McLaughlin PJ, West DC, McFadyen IR, Johnson PM. Endothelial cell activation by tumour necrosis factor-alpha (TNF-alpha) and the development of pre-eclampsia. Clin Exp Immunol. 1994; 98: 110-4.

72.Meisser A, Cameo P, Islami D, Campana A & Bischof P. (1999): Effects of interleukin-6 (IL-6) on cytotrophoblastic cells. Molecular Human Reproduction, 5:1055–1058.

73.Murotsuki J, Challis JRG, Han VKM, Fraher J and Gagnon R (1997) Chronic fetal placental embolization and hypoxaemia cause hypertension and myocardial hypertrophy in fetal sheep American Journal of Physiology 272 R201–R207

74.Myatt L, Brewer A, Brockman D (1991) The action of nitric oxide in the perfused human fetal-placental circulation. Am J Obstet Gynecol 164:687-692.

75.Nakatsuka M, Habara T, Noguchi S, Konishi H, Kudo T (2003). Increased plasma adrenomedullin in women with recurrent pregnancy loss. Obstetrics and gynecology;102:319-24.

76.Nawroth P.P, Stern D.M. (1986): " Modulation of endothelial cells hemostatic properties by tumor necrosis factor." Journal of Experimental Medicine 163:740-745.

77.Odegard RA, Vatten LJ, Nilsen ST, Salvesen KA, Vefring H, Austgulen R. Umbilical cord plasma interlukin-6 and fetal growth restriction in preeclampsia: a prospective study in Norway. Obstetrics & gynecology 2001;98:289-94.

78.Opsjon SL, Austgulen R, Waage A. Interleukin-1, interleukin-6 and tumor necrosis factor at delivery in preeclamptic disorders. Acta Obstet Gynecol Scand 1995;74:19–26.

79.Phillips TA, Ni J and Hunt JS (2001): Death-inducing tumour necrosis factor (TNF) superfamily ligands and receptors are transcribed in human placentae, cytotrophoblasts, placental macrophages and placental cell lines. Placenta 22, 663–672.

80.Pijnenborg R, McLaughlin PJ, Vercruysse L, Hanssens M, Johnson PM, Keith Jr JC, et al. (1998): Immunolocalization of tumour necrosis factor-alpha (TNF alpha) in the placental bed of normotensive and hypertensive human pregnancies. Placenta; 19:231–9.

81.Raymond W, Redline (2007) Villitis of unknown etiology: noninfectious chronic villitis in the placenta. Human Pathology 38:1439-1446.

82.Rivera-Quinoes, C., McGavern, D., Schmelzer, J.D., Hunter, S.F. Low, P.A., Rodriguez, M., 1998. Absence of neurological deficits following extensive demyelination in a class I-deficient murine model of multiple sclerosis. Nat. Med. 4, 187.

83.Roberts JM and Cooper DW. (2001): Pathogenesis and genetics of preeclampsia. Lancet, 357:53–6.

84.Salafia CM, Vintzileos AM, Silberman L, Bantham KF, Vogel CA (1992) Placental pathology of idiopathic intrauterine growth retardation at term. Am J Perinatol 9(3):179-84.

85.Schiff E, Friedman SA, Baumann P, Sibai BM, Romero R. Tumor necrosis factor-alpha in pregnancies associated with preeclampsia or small-for-gestational-age newborns. Am J Obstet Gynecol 1994;170:1224–9.

86.Sebire NJ (2003) Umbilical artery Doppler revisited: pathophysiology of changes in intrauterine growth restriction revealed Ultrasound. Obstetrics & Gynecology, 21:419-422.

87.Sebire NJ, Goldin RD, Regan L (2001) Histomorphological evidence for chronic vasoconstriction of placental stem vessels in pregnancies with intrauterine growth restriction and abnormal umbilical artery. Doppler velocimetry indices   J Pathol 195:19A.

88.Sebire NJ, Talbert D (2002) The role of intraplacental vascular smooth muscle in the dynamic placenta: a conceptual framework for understanding uteroplacental disease. Med Hypotheses 58:347-351.

89. Seremak-Mrozikiewicz A, Dubiel M, Drews K, Gudmundsson S, Mrozikiewicz PM (2008): TNF-alpha gene polymorphism and fetal Doppler velocimetry in intrauterine growth restriction. Neuro Endocrinol Lett. 29(4):493-9.

90. Simmons DG, Fortier AL, Cross JC (2007) Diverse subtypes and developmental origins of trophoblast giant cells in the mouse placenta. Dev Biol. 304:567-578.

91. Smith SC, Baker PN (1999): Placental apoptosis is increased in post-term pregnancies. Br J Obstet Gynecol106:861-862.

92. Smith SC, Baker PN, Symonds EM (1997): Increased placental apoptosis in intrauterine growth restriction. Am J Obstet Gynecol 177:1395-1401.

93. Smith SC, Baker PN, Symonds EM (1997): Placental apoptosis in normal human pregnancy. Am J Obstet Gynecol, 177:57–65.

94. Stallmach T, Hebisch G, Joller H, Kolditz P, Engelmann M (1995) Expression pattern of cytokines in the different compartments of the feto-maternal unit under various conditions. Reprod Fertil Dev. 7:1573-80.

95. Stonestreet BS, Widness JA, Berard DJ. Circulatory and metabolic effects of hypoxia in the hyperinsulinemic ovine fetus. Pediatr Res. 1995 Jul;38(1):67-75.

96. Street ME, Seghini P, Fieni S, Ziveri MA, Volta C, Martorana D, Viani I, Gramellini D, Bernasconi S. Changes in interleukin-6 and IGF system and their relationships in placenta and cord blood in newborns with fetal growth restriction compared with controls. European Journal of Endocrinology 2006;155:(4):567–574.

97. Sur M, AlArdati H, Ross C and Alowami S (2007): TdT expression in Merkel cell carcinoma: potential diagnostic pitfall with blastic hematological malignancies and expanded immunohistochemical analysis. Modern Pathology 20, 1113–1120

98. Tinkanen H, Rorarius M, Metsä-Ketelä T. Catecholamine concentrations in venous plasma and cerebrospinal fluid in normal and complicated pregnancy. Gynecol Obstet Invest. 1993;35(1):7-11.

99. Tomasa SZ, Rojeb D, Prusacc IK, Tadind I, Capkunb V (2010) Morphological characteristics of placentae associated with idiopathic intrauterine growth retardation: a clinicopathologic study. European Journal of Obstetrics & Gynecology and Reproductive Biology. 152(1):39-43.

100. Tosun M, Celik H, Avci B, Yavuz E, Alper T, Malatyalioğlu E (2010): Maternal and umbilical serum levels of interleukin-6, interleukin-8, and tumor

necrosis factor-alpha in normal pregnancies and in pregnancies complicated by preeclampsia. J Matern Fetal Neonatal Med. 23(8):880-6.

101.   Upton P.D., Austin C., Taylor G.M., Nandha K.A., Clark A.J.L., Ghatei M.A., Bloom S.R., Smith D.M. Expression of adrenomedullin (ADM) and its binding sites in the rat uterus: increased number of binding sites and ADM messenger ribonucleic acid in 20-day pregnant rats compared with nonpregnant rats. Endocrinology. 1997;138:2508–2514.

102.   Villar J, Carroli G, Wojdyla D, Abalos E, Giordano D et al. (2005): for the World Health Organization Antenatal Care.

103.   Villar J. Ezcurra EJ, Gurtner de la Fuente V & Campodonico L (1994): Pre-term delivery syndrome: the unmet need In New Perspectives for the effective treatment of pre-term labor: an international consensus. Research and Clinical Forums; 16, pp 9-38.

104.   Vince G, Shorter S, Starkey P, Humphreys J, Clover L, Wilkins T, et al. (1992): Localization of tumour necrosis factor production in cells at the materno/fetal interface in human pregnancy. Clin Exp Immunol. 88:174–80.

105.   W.H.O. (1995) Expert Committee Report: Physical status: the use and interpretation of anthropometry. Technical Report Series 854. Geneva: World Health Organization.

106.   Wang L, Du F, Wang X (2008) TNF-alpha induces two distinct caspase-8 activation pathways. Cell. 133:693-703.

107.   Williams RL, Creasy RK, Cunningham GC, Hawes WE, Norris FD, Tashiro M (1982) Fetal growth and perinatal viability in California. Obstet Gynecol 59:624-32.

108.   Wride MA, Sanders EJ (1995) Potential roles for tumour necrosis factor-α during embryonic development. Anat Embryol. 191:1-10.

109.   Yang Y, Yelavarthi KK, Chen HL, Pace JL, Terranova PF, Hunt JS (1993) Molecular, biochemical, and functional characteristics of tumor necrosis factor-alpha produced by human placental cytotrophoblastic cells. J Immunol. 150:5614-24.

110.   Yelavarthi KK, Hunt JS (1993) Analysis of p60 and p80 tumor necrosis factor-a receptor messenger RNA and protein in human placentas. Am J Pathol. 143:1131-1141.

111.   Yu X, Wang L, Yan C, Li X (2007) Expression and localization of tumor necrosis factor receptor 1 protein in the chorionic villi in early normal and spontaneous abortion. Europ J Obstet Gynecol Reprod Biol. 132: 58-63.

112.   Yui J, Garcia-Lloret M, Wegmann TG, Guilbert LJ (1994) Cytotoxicity of tumor necrosis factor alpha and gamma-interferon against primary human placental trophoblasts. Placenta. 15:819-835.

113.   Zhang W, Wang L, Liu L. The study of levels of norepinephrine and dopamine-beta-hydroxylase in patients with pregnancy-induced hypertension. Zhonghua Yi Xue Za Zhi (Taipei). 2001 Jun;64(6):351-6.

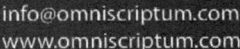

Printed by Books on Demand GmbH, Norderstedt / Germany